I0150733

The Mule

A Treatise On The Breeding, Training, And Uses To Which He May Be Put

Harvey Riley

Contents

THE MULE
A TREATISE ON THE BREEDING,
TRAINING, AND USES TO WHICH HE
MAY BE PUT
BY
Harvey Riley

PREFACE.

There is no more useful or willing animal than the Mule. And perhaps there is no other animal so much abused, or so little cared for. Popular opinion of his nature has not been favorable; and he has had to plod and work through life against the prejudices of the ignorant. Still, he has been the great friend of man, in war and in peace serving him well and faithfully. If he could tell man what he most needed it would be kind treatment. We all know how much can be done to improve the condition and advance the comfort of this animal; and he is a true friend of humanity who does what he can for his benefit. My object in writing this book was to do what I could toward working out a much needed reform in the breeding, care, and treatment of these animals. Let me ask that what I have said in regard to the value of kind treatment be carefully read and followed. I have had thirty years' experience in the use of this animal, and during that time have made his nature a study. The result of that study is, that humanity as well as economy will be best served by kindness.

It has indeed seemed to me that the Government might make a great saving every year by employing only such teamsters and wagon-masters as had been thoroughly instructed in the treatment and management of animals, and were in every way qualified to perform their duties properly. Indeed, it would seem only reasonable not to trust a man with a valuable team of animals, or perhaps a train, until he had been thoroughly instructed in their use, and had received a certificate of capacity from the Quartermaster's Department. If this were done, it would go far to establish a system that would check that great destruction of animal life which costs the Government so heavy a sum every year.

H.R.

WASHINGTON, D.C., *April 12, 1867*.

NOTE.

I have, in another part of this work, spoken of the mule as being free from splint. Perhaps I should have said that I had never seen one that had it, notwithstanding the number I have had to do with. There are, I know, persons who assert that they have seen mules that had it. I ought to mention here, also, by way of correction, that there is another ailment the mule does not have in common with the horse, and that is quarter-crack. The same cause that keeps them from having quarter-crack preserves them from splint--the want of front action.

A great many persons insist that a mule has no marrow in the bones of his legs. This is a very singular error. The bone of the mule's leg has a cavity, and is as well filled with marrow as the horse's. It also varies in just the same proportion as in the horse's leg. The feet of some mules, however, will crack and split, but in most cases it is the result of bad shoeing. It at times occurs from a lack of moisture to the foot; and is seen among mules used in cities, where there are no facilities for driving them into running water every day, to soften the feet and keep them moist.

CHAPTER I. HOW MULES SHOULD BE TREATED IN BREAKING.

I have long had it in contemplation to write something concerning the mule, in the hope that it might be of benefit to those who had to deal with him, as well in as out of the army, and make them better acquainted with his habits and usefulness. The patient, plodding mule is indeed an animal that has served us well in the army, and done a great amount of good for humanity during the late war. He was in truth a necessity to the army and the Government, and performed a most important part in supplying our army in the field. That he will perform an equally important part in the future movements of our army is equally clear, and should not be lost sight of by the Government. It has seemed to me somewhat strange, then, that so little should have been written concerning him, and so little pains taken to improve his quality. I have noticed in the army that those who had most to do with him were the least acquainted with his habits, and took the least pains to study his disposition, or to ascertain by proper means how he could be made the most useful. The Government might have saved hundreds of thousands of dollars, if, when the war began, there had been a proper understanding of this animal among its employees.

Probably no animal has been the subject of more cruel and brutal treatment than the mule, and it is safe to say that no animal ever performed his part better, not even the horse. In breaking the mule, most persons are apt to get out of patience with him. I have got out of patience with him myself. But patience is the great essential in breaking, and in the use of it you will find that you get along much better. The mule is an unnatural animal, and hence more timid of man than the horse; and yet he is tractable, and capable of being taught to understand what you want him to do. And when he understands what you want, and has gained your confidence, you will, if you treat him kindly, have little trouble in making him perform his duty.

In commencing to break the mule, take hold of him gently, and talk to him kindly. Don't spring at him, as if he were a tiger you were in dread of. Don't yell at him; don't jerk him; don't strike him with a club, as is too often done; don't get excited at his jumping and kicking. Approach and handle him the same as you would an animal already broken, and through kindness you will, in less than a week, have your mule more tractable, better broken, and kinder than you would in a month, had you used the whip. Mules, with very few exceptions, are born kickers. Breed them as you will, the moment they are able to stand up, and you put your hand on them, they will kick. It is, indeed, their natural means of defence, and they resort to it through the force of instinct. In commencing to break them, then, kicking is the first thing to guard against and overcome. The young mule kicks because he is afraid of a man. He has seen those intrusted with their care beat and abuse the older ones, and be very naturally fears the same treatment as soon as a man approaches him. Most persons intrusted with the care of these young and green mules have not had experience enough with them to know that this defect of kicking is soonest remedied by kind treatment. Careful study of the animal's nature and long experience with the animal have taught me that, in breaking the mule, whipping and harsh treatment almost invariably make him a worse kicker. They certainly make him more timid and afraid of you. And just as long as you fight a young mule and keep him afraid of you, just so long will you be in danger of his kicking you. You must convince him through kindness that you are not going to hurt or punish him. And the sooner you do this, the sooner you are out of danger from his feet.

It may at times become necessary to correct the mule before he is subdued; but before doing so he should be well bridle or halter-broken, and also used to harness. He should also be made to know what you are whipping him for. In harnessing up a mule that will kick or strike with the forefeet, get a rope, or, as we term it in the army, a lariat. Throw, or put the noose of this over his head, taking care at the same time that it be done so that the noose does not choke him; then get the mule on the near side of a wagon, put the end of the lariat through the space between the spokes of the fore wheel, then pull the end through so that you can walk back with it to the hinder wheel (taking care to keep it tight), then pass it through the same, and pull the mule close to the wagon. In this position you can bridle and harness him without fear of being crippled. In putting the rope through the above places, it should be put through the wheels, so as to bring it as high as the mule's breast in front, and flanks in the rear. In making them fast in this way, they frequently kick until they get over the rope, or lariat; hence the necessity of keeping it as high up as possible. If you chance upon a mule so wild that you cannot handle him in this way, put a noose of the lariat in the mule's mouth, and let the eye, or the part where you put the end of the lariat through, be so as to form another noose. Set this directly at the root of the mule's ear, pull it tight on him, taking care to keep the noose in the same place. But when you get it pulled tight enough, let some one hold the end of the lariat, and, my word for it, you will bridle the mule without much further trouble.

In hitching the mule to a wagon, if he be wild or vicious, keep the lariat the same as I have described until you get him hitched up, then slack it gently, as nearly all mules will buck or jump stiff-legged as soon as you ease up the lariat; and be careful not to pull the rope too tight when first put on, as by so doing you might split the mule's mouth. Let me say here that I have broken thousands of four and six-mule teams that not one of the animals had ever had a strap of harness on when I began with them, and I have driven six-mule teams for years on the frontier, but I have yet to see the first team of unbroken mules that could be driven with any degree of certainty. I do not mean to say that they cannot be got along the road; but I regard it no driving worthy of the name when a driver cannot get his team to any place where he may desire to go in a reasonable time--and this he cannot do with unbroken mules. With green or unbroken mules, you must chase or herd them along without the whip, until you get them to know that you want them to pull in a wagon. When you have got them in a wagon, pull their heads round in the direction you want them to go; then convince them by your kindness that you are not going to abuse them, and in twelve days' careful handling you will be able to drive them any way you please.

In bridling the young mule, it is necessary to have a bit that will not injure the animal's mouth. Hundreds of mules belonging to the Government are, in a measure, ruined by using a bridle bit that is not much thicker than the wire used by the telegraph. I do not mean by this that the bridle bit used by the Government in its blind bridles is not well adapted to the purpose. If properly made and properly used, it is. Nor do I think any board of officers could have gotten up or devised a better harness and wagon for army purposes than those made in conformity with the decision of the board of officers that recommended the harness and wagon now used. The trouble with a great many of the bits is, that they are not made up to the regulations, and are too thin. And this bit, when the animal's head is reined up too tight, as army teamsters are very likely to do, is sure to work a sore mouth.

There are few things in breaking the mule that should be so carefully guarded against as this. For as soon as the animal gets a sore mouth, he cannot eat well, and becomes fretful; then he cannot drink well, and as his mouth keeps splitting up on the sides, he soon gets so that he cannot keep water in it, and every swallow he attempts to take, the water will spirt out of the sides, just above the bit. As soon as the mule finds that he cannot drink without this trouble, he very naturally pushes his nose into the water above where his mouth is split, and drinks until the want of breath forces him to stop, although he has not had sufficient water. The animal, of course, throws up its head, and the stupid teamster, as a general thing, drives the mule away from the water with his thirst about half satisfied.

Mules with their mouths split in this way are not fit to be used in the teams, and the sooner they are taken out and cured the better for the army and the Government. I have frequently seen Government trains detained several minutes, block the road, and throw the train into disorder, in order to give a mule with a split mouth time to drink. In making up teams for a train, I invariably leave out all mules whose mouths are not in a sound state, and this I do without regard to the kind or quality of the animal. But the mule's mouth can be saved from the condition I have referred to, if the bit be made in a proper manner.

The bit should be one inch and seven-eighths round, and five inches in the draw, or between the rings. It should also have a sweep of one quarter of an inch to the five inches long. I refer now to the bit for the blind bridle. With a bit of this kind it is almost impossible to injure the mule's mouth, unless he is very young, and it cannot be done then if the animal is handled with proper care.

There is another matter in regard to harnessing the mule which I deem worthy of notice here. Government teamsters, as a general thing, like to see a mule's head reined tightly up. I confess that, with all my experience, I have never seen the benefit there was to be derived from this. I always found that the mule worked better when allowed to carry his head and neck in a natural position. When not reined up at all, he will do more work, out-pull, and wear out the one that is. At present, nearly all the Government mule-teams are reined up, and worked with a single rein. This is the old Virginia way of driving mules. It used to be said that any negro knew enough to drive mules. I fear the Government has too long acted on that idea.

I never heard but one reason given for reining the heads of a mule-team up tight, and that was, that it made the animals look better.

The next thing requiring particular attention is the harnessing. During the war it became customary to cut the drawing-chains, or, as some call them, the trace-chains. The object of this was, to bring the mule close up to his work. The theory was taken from the strings of horses used in drawing railroad cars through cities. Horses that are used for hauling cars in this manner are generally fed morning, noon, and night; and are able to get out of the way of a swingle-tree, should it be let down so low as to work on the brakes, as it did too frequently in the army. Besides, the coupling of the car, or the part they attach the horse to, is two-thirds the height of a common-sized animal, which, it will be seen at a glance, is enough to keep the swingle-tree off his heels. Now, the tongue of a Government wagon is a very different thing. In its proper condition, it is about on an average height with the mule's hocks; and, especially during the last two years of the war, it was customary to pull the mule so close up to the swingle-tree that his hocks would touch it. The result of hitching in this manner is, that the mule is continually trying to keep out of the way of the swingle-tree, and, finding that he cannot succeed, he becomes discouraged. And as soon as he does this he will lag behind; and as he gets sore from this continual banging, he will spread his hind legs and try to avoid the blows; and, in doing this, he forgets his business and becomes irritable. This excites the teamster, and, in ninety-nine cases out of a hundred, he will beat and punish the animal cruelly, expecting thereby to cure him of the trouble. But, instead of pacifying the mule, he will only make him worse, which should, under no circumstances, be done. The proper course to pursue, and I say so from long experience, is to stop the team at once, and let all the traces out to a length that will allow the swingle-tree to swing half way between the hock and the heel of the hoof. In other words, give him room enough to step, between the collar and swingle-tree, so that the swingle-tree cannot touch his legs when walking at his longest stride. If the above rule be followed, the animal will not be apt to touch the swingle-tree. Indeed, it will not be apt to touch him, unless he be lazy; and, in that case, the sooner you get another mule the better. I say this because one lazy mule will spoil a good team, invariably. A lazy mule will be kept up to his work with a whip, you will say; but, in whipping a lazy animal, you keep the others in such a state of excitement that they are certain to get poor and valueless.

There is another advantage in having the drawing-chains worked at the length I have described. It is this: The officers that formed the board that recommended the drawing-chain, also recommended a number of large links on one end of the chain, so that it could be made longer or shorter, as desired. If made in conformity with the recommendation of that board of officers, it can be let out so as to fit the largest sized mule, and can be taken up to fit the shortest. When I say this, I mean to include such animals as are received according to the standard of the Quarter-master-General's department.

CHAPTER II. THE DISADVANTAGES OF WORKING MULES THAT ARE TOO YOUNG.

A great many of the mules purchased by the Government during the war were entirely too young for use. This was particularly so in the West, where both contractor and inspector seemed anxious only to get the greatest number they could on the hands of the Government, without respect to age or quality. I have harnessed, or rather tried to harness, mules during the war, that were so young and small that you could not get collars small enough to fit them. As to the harness, they were almost buried in it. A great many of these small mules were but two years old. These animals were of no use to the Government for a long time. Indeed, the inspector might just as well have given his certificate for a lot of milk cows, so far as they added to our force of transportation. Another source of trouble has been caused through a mistaken opinion as to what a young mule could do, and how he ought to be fed. Employers and others, who had young mules under their charge during the war, had, as a general thing, surplus forage on hand. When they were in a place where nine pounds of grain could be procured, and fourteen of hay, the full allowance was purchased. The surplus resulting from this attracted notice, and many wondered why it was that the Government did not reduce the forage on the mule. These persons did not for a moment suspect, or imagine, that a three year old mule has so many loose teeth in his mouth as to be hardly able to crack a grain of corn, or masticate his oats.

Another point in that case is this: at three years old, a mule is in a worse condition, generally, than he is at any other period in life. At three, he is more subject to distemper, sore eyes, and inflammation of all parts of the head and body. He becomes quite weak from not being able to eat, gets loose and gaunt, and is at that time more subject and more apt to take contagious diseases than at any other change he may go through. There is but one sure way to remedy this evil. Do not buy three year old mules to put to work that it requires a five or six year old mule to perform. Six three year old mules are just about as fit to travel fifteen miles per day, with an army wagon loaded with twenty-five hundred and their forage, as a boy, six years of age, is fit to do a man's work. During the first twelve months of the war, I had charge of one hundred and six mule-teams, and I noticed in particular, that not one solitary mule as high as six years old gave out on the trips that I made with the teams. I also noticed that, on most occasions, the three year olds gave out, or became so leg-weary that they could scarce walk out of the way of the swingle-tree, whereas those of four and upward would be bright and brisk, and able to eat their forage when they came to camp. The three year old mules would lie down and not eat a bite, through sheer exhaustion. I also noticed that nearly all the three year old mules that went to Utah, in 1857, froze to death that winter, while those whose ages varied from four, and up to ten, stood the winter and came out in the spring in good working condition. In August, 1855, I drove a six-mule team to Fort Riley, in Kansas Territory, from Fort Leavenworth, on the Missouri River, loaded with twelve sacks of grain. It took us thirteen days to make the trip. When we reached Fort Riley there were not fifty mules, in the train of one hundred and fifty, that would have sold at public sale for thirty dollars, and a great many gave out on account of being too young and the want of proper treatment. In the fall of 1860, I drove a six-mule team, loaded with thirty hundred weight, twenty-five days' rations for myself and another man, and twelve days' storage for the team, being allowed twelve pounds to each mule per day. I drove this team to Fort Laramie, in Nebraska Territory, and from there to Fort Leavenworth, on the Missouri River. I made the drive there and back in thirty-eight days, and laid over two and a half days out of that. The distance travelled was twelve hundred and thirty-six miles. After a rest of two days, I started with the same team, and drove to Fort Scott, in Kansas Territory, in five days, a distance of one hundred and twenty miles. I went with Harney's command, and, for the most part of the time, had no hay, and was forced to subsist our animals on dry prairie grass, and had a poor supply of even that. Notwithstanding this, I do not believe that any mule in the team lost as much as ten pounds of flesh. Each of these mules, let me say, was upward of five years old.

In 1858, I took a train of mules to Camp Floyd, in Utah, forty-eight miles south of Salt Lake City; During the march there were days and nights that I could not get a drop of water for the animals. The young mules, three and four years old, gave out from sheer exhaustion; while the older ones kept up, and had to draw the wagons along. Now, there are many purposes to which a young mule may be put with advantage; but they are altogether unfit for army purposes, and the sooner the Government stops using them, the better.

When they are purchased for army use, they are almost sure to be put into a train, and turned over to the tender mercies of some teamster, who knows nothing whatever about the character of the animal. And here let me say that thousands of the best mules in the army, during the war, were ruined and made useless to the Government on account of the incompetency and ignorance of the wagon-masters and teamsters who had to deal with them. Persons who own private teams and horses are generally particular to know the character of the person who takes care of them, and to ascertain that he knows his business. Is he a good driver? Is he a good groom? Is he careful in feeding and watering? These are the questions that are asked; and if he has not these qualities he will not do. But a teamster in the army has none of these questions put to him. No; he is intrusted with a valuable team, and expected to take proper care of it when he has not the first qualification to do so. If he is asked a question at all, it is merely if he has ever driven a team before. If he answer in the affirmative, and there are any vacancies, he is employed at once, though he may not know how to lead a mule by the head properly. This is not alone the case with teamsters. I have known wagon-masters who really did not know how to straighten out a six-mule team, or, indeed, put the harness on them properly. And yet the wagon-master has almost complete power over the train. It will be readily seen from this, how much valuable property may be destroyed by placing incompetent men in such places. Wagon-masters, it seems to me, should not be allowed, under any circumstances, to have or take charge of a train of animals of any kind until they are thoroughly competent to handle, harness, and drive a six-animal team.

There is another matter which needs essential improvement. I refer now to the men who are placed as superintendents over our Government corrals and depots for animals. Many of these men know little of either the horse or the mule, and are almost entirely ignorant of what is necessary for transportation. A superintendent should have a thorough knowledge of the character and capacity of all kinds of animals necessary for a good team. He should know at sight the age and weight of animals, should be able to tell the most suitable place for different animals in a team, and where each would be of the most service. He should know all parts of his wagon and harness at a glance, be able to take each portion apart and put them together again, each in its proper shape and place, and, above all, he should have practical experience with all kinds of animals that are used in the army. This is especially necessary during war.

CHAPTER III. COLOR, CHARACTER, AND PECULIARITIES OF MULES.

After being in command of the upper corral, I was ordered, on the 7th of September, 1864, to take charge of the Eastern Branch Wagon Park, Washington. There were at that time in the park twenty-one six-mule trains. Each train had one hundred and fifty mules and two horses attached. There were times, however, when we had as many as forty-two trains of six-mule teams, with thirty men attached to each train. In a year from the above date we handled upward of seventy-four thousand mules, each and every one passing under my inspection and through my hands.

In handling this large number of animals, I aimed to ascertain which was the best, the hardest, and the most durable color for a mule. I did this because great importance has been attached by many to the color of these animals. Indeed, some of our officers have made it a distinguishing feature. But color, I am satisfied, is no criterion to judge by. There is an exception to this, perhaps, in the cream-colored mule. In most cases, these cream-colored mules are apt to be soft, and they also lack strength. This is particularly so with those that take after the mare, and have manes and tails of the same color. Those that take after the jack generally have black stripes round their legs, black manes and tails, and black stripes down their backs and across their shoulders, and are more hardy and better animals. I have frequently seen men, in purchasing a lot of mules, select those of a certain color, fancying that they were the hardiest, and yet the animals would be widely different in their working qualities. You may take a black mule, black mane, black hair in his ears, black at the flank, between the hips or thighs, and black under the belly, and put him alongside of a similar sized mule, marked as I have described above, say light, or what is called mealy-colored, on each of the above-mentioned parts, put them in the same condition and flesh, of similar age and soundness, and, in many cases, the mule with the light-colored parts will wear the other out.

It is very different with the white mule. He is generally soft, and can stand but little hardship. I refer particularly to those that have a white skin. Next to the white and cream, we have the iron-grey mule. This color generally indicates a hardy mule. We have now twelve teams of iron-gray mules in the park, which have been doing hard work every day since July, 1865; it is now January, 1866. Only one of these mules has become unfit for service, and that one was injured by being kicked by his mate. All our other teams have had more or less animals made unfit for service and exchanged.

In speaking of the color of mules, it must not be inferred that there are no mules that are all of a color that are not hardy and capable of endurance. I have had some, whose color did not vary from head to foot, that were capable of great endurance. But in most cases, if kept steadily at work from the time they were three years old until they were eight or ten, they generally gave out in some part, and became an expense instead of profit.

Various opinions are held as to what the mule can be made to do under the saddle, many persons asserting that in crossing the plains he can be made to perform almost equal to the horse. This is true on the prairie. But there he works with every advantage over the horse. In 1858, I rode a mule from Cedar Valley, forty-eight miles north of Salt Lake City, to Fort Leavenworth, Kansas, a distance of nearly fourteen hundred miles. Starting from Cedar Valley on the 22d of October, I reached Fort Leavenworth on the 31st of December. At the end of the journey the animal was completely worn down.

In this condition I put her into Fleming's livery stable, in Leavenworth City, and was asked if she was perfectly gentle. One would suppose that, in such a condition, she would naturally be so. I assured the hostler that she was; that I had ridden her nearly a year, and never knew her to kick. That same morning, when the hostler went to feed her, she suddenly became vicious, and kicked him very severely. She was then about twelve years old. I have since thought that when a mule gets perfectly gentle he is unfit for service.

Proprietors of omnibuses, stage lines, and city railroads have, in many cases, tried to work mules, as a matter of economy; but, as a general thing, the experiment proved a failure, and they gave it up and returned to horses. The great reason for this failure was, that the persons placed in charge of them knew nothing of their disposition, and lacked that experience in handling them which is so necessary to success. But it must be admitted that, as a general thing, they are not well adapted for road or city purposes, no matter how much you may understand driving and handling them.

The mule may be made to do good service on the prairies, in supplying our army, in towing canal boats in hauling cars inside of coal mines-- these are his proper places, where he can jog along and take his own time, patiently. Work of this kind would, however, in nearly all cases, break down the spirit of the horse, and render him useless in a very short time.

I have seen it asserted that there were mules that had been known to trot in harness in three minutes. In all my experience, I have never seen any thing of the kind, and do not believe the mule ever existed that could do it. It is a remarkably good road horse that will do this, and I have never yet seen a mule that could compare for speed with a good roadster. I have driven mules, single and double, night and day, from two to ten in a team, and have handled them in every way that it is possible to handle them, and have in my charge at this time two hundred of the best mule teams in the world, and there is not a span among them that could be forced over the road in four minutes. It is true of the mule that he will stand more abuse, more beating, more straining and constant dogging at him than any other animal used in a team. But all the work you can get out of him, over and above an ordinary day's work, you have to work as hard as he does to accomplish.

Some curious facts have come under my knowledge as to what the mule can endure. These facts also illustrate what can be done with the animal by persons thoroughly acquainted with his character. While on the plains, I have known Kiowa and Camanche Indians to break into our pickets during the night, and steal mules that had been pronounced completely broken down by white men. And these mules they have ridden sixty and sixty-five miles of a single night. How these Indians managed to do this, I never could tell. I have repeatedly seen Mexicans mount mules that our men had pronounced unfit for further service, and ride them twenty and twenty-five miles without stopping. I do not mention this to show that a Mexican can do more with the mule than an American. He cannot. And yet there seems to be some sort of fellow-feeling between these Mexicans and the mule. One seems to understand the other completely; and in disposition there is very little difference. And yet the Mexican is so brutish in dealing with animals, that I never allowed one of them to drive a Government team for me. Indeed, a low Mexican does not seem disposed to work for a man who will not allow him full latitude in the abuse of animals.

Packing Mules.--The Mexican is a better packer than the American. He has had more experience, and understands all its details better than any other man. Some of our United States officers have tried to improve on the experience of the Greaser, and have made what they called an improvement on the Mexican pack-saddle. But all the attempts at improvement have been utter failures. The ranchero, on the Pacific side of the Sierra Nevadas, is also a good packer; and he can beat the Mexican lassoing cattle. But he is the only man in the United States who can. The reason for this is, that they went into that country when very young, and improved on the Mexican, by having cattle, mules, and horses round them all the time, and being continually catching them for the purpose of branding and marking.

There is, in Old as well as New Mexico, a class of mules that are known to us as Spanish, or Mexican mules. These mules are not large, but for endurance they are very superior, and, in my opinion, cannot be excelled. I am not saying too much when I assert, that I have seen nothing in the United States that could compare with them. They can, apparently, stand any amount of starvation and abuse. I have had three Spanish mules in a train of twenty-five six-mule teams, and starting from Fort Leavenworth, Kansas, on Colonel (since General) Sumner's expedition, in 1857, have travelled to Walnut Creek, on the Santa Fe route, a distance of three hundred miles, in nine days. And this in the month of August. The usual effects of hard driving, I noticed, showed but very little on them. I noticed also, along the march, that with a halt of less than three hours, feeding on grass that was only tolerably thick, they will fill up better and look in better condition for resuming the march, than one of our American mules that had rested five hours, and had the same forage. The breed, of course, has something to do with this. But the animal is smaller, more compact than our mules, and, of course, it takes less to fill him up. It stands to reason, that a mule with a body half as large as a hogshead cannot satisfy his hunger in the time it would take a small one. This is the secret of small mules outlasting large ones on the prairies. It takes the large one so long to find enough to eat, when the grass is scanty, that he has not time enough for rest and recuperation. I often found them leaving camp, in the morning, quite as hungry and discouraged as they were when we halted the previous evening. With the small mule it is different. He gets enough to eat, quick, and has time to rest and refresh himself. The Spanish or Mexican mule, however, is better as a pack animal, than for a team. They are vicious, hard to break, and two-thirds of them kick.

In looking over a book, with the title of "Domestic Animals," I notice that the author, Mr. R.L. Allen, has copied from the official report of the Agricultural Committee of South Carolina, and asserts that a mule is fit for service sooner than a horse. This is not true; and to prove that it is not, I will give what I consider to be ample proof. In the first place, a mule at three years old is just as much and even more of a colt than a horse is. And he is as much out of condition, on account of cutting teeth, distemper, and other colt ailments, as it is possible to be. Get a three year old mule tired and fatigued, and in nine cases out of ten he will get so discouraged that it will be next to impossible to get him home or into camp. A horse colt, if able to travel at all, will work his way home cheerfully; but the young mule will sulk, and in many instances will not move an inch while life lasts. An honest horse will try to help himself, and do all he can for you, especially if you treat him kindly. The mule colt will, just as likely as not, do all he can to make it inconvenient for you and him.

To show of how little service three year old mules are to the Government, I will give the number handled by me during part of 1864 and 1865.

On the 1st of September, 1864, I had charge of five thousand and eighty-two mules; and during the same month I received two thousand two hundred and ten, and issued to the Armies of the Potomac, the James, and the Shenandoah, three thousand five hundred and seventy-one, which left us on hand, on the 1st of October, three thousand seven hundred and twenty-one. During the month of October we received only nine hundred and eighty, and issued two thousand five hundred and thirty, which left us on hand, on the 1st of November, two thousand one hundred and seventy-one. During November we received two thousand one hundred and eighty-six, and issued to the army one thousand seven hundred and fifty-seven, which left us on hand, on the 1st of December, two thousand four hundred and thirty mules. Now mark the deaths.

During the month of September, 1864, there died in the corral fifteen mules. In October, six died. In November, three; and in December, eight. They were all two and three years old.

On the 1st of May, 1865, we had on hand four thousand and twelve head, and received, during the same month, seven thousand nine hundred and fifty-eight. We issued, during the same month, fifteen thousand five hundred and sixty-three, leaving us on hand, on the 1st of June, six thousand four hundred and eighty-seven. During this month we received seven thousand nine hundred and fifty-one, and issued eleven thousand nine hundred and fifteen. Our mules during these months were sent out to be herded, and the total number of deaths during the time was twenty-four. But two of them were over four years old. Now, it occurs to me that it would be a great saving to the Government not to purchase any mules under four years old. This statement of deaths at the corral is as nothing when compared with the number of deaths of young mules in the field. It is, in fact, well established that fully two-thirds of the deaths in the field are of young animals under three years of age. This waste of animal life carries with it an expense it would be difficult to estimate, but which a remedy might easily be found for.

Now, it is well known that when a mule has reached the age of four years, you will have very little trouble with him, so far as sickness and disease are concerned. Besides, at the age of four he is able to work, and work well; and he also understands better what you want him to do.

The committee appointed to report on this subject say many mules have been lost by feeding on cut straw and corn meal. This is something entirely new to me; and I am of opinion that more Government mules die because they do not get enough of this straw and meal. The same committee say, also, that in no instance have they known them to be inflicted with disease other than inflammation of the intestines, caused by exposure. I only wish that the members of that committee could have had access to the affidavits in the Quartermaster-General's department-- they would then have satisfied themselves that thousands of Government mules have died with almost every disease the horse is subject to. And I do not see why they should not be liable to the same diseases, since they derive life and animation from the horse. The mule that breeds closest after the jack, and is marked like him, is the hardiest, can stand fatigue the best, and is less liable to those diseases common to the horse; while those which breed close after the mare, and have no marks of the jack about them, are liable to all of them.

In the beginning of this chapter I spoke of the color of mules. I will, in closing, make a few more remarks on that subject, which may interest the reader. We have now at work three dun-colored mules, that were transferred to the Army of the Potomac in 1862, and that went through all the campaigns of that army, and were transferred back to us in June, 1865. They had been steadily at work, and yet were in good condition, hardy, and bright, when they were turned in. These mules have a black stripe across their shoulders, down their backs, and are what is called "dark-colored duns." We also have the only full team that has gone through all the campaigns of the Army of the Potomac. It was fitted up at Annapolis, Md., in September, 1861, under Captain Santelle, A.Q.M. They are now in fine condition, and equal to any thing we have in the corral. The leaders are very fine animals. They are fourteen hands high, one weighing eight hundred, and the other eight hundred and forty-five pounds. One of the middle leaders weighs nine hundred, the other nine hundred and forty-seven pounds, and fourteen hands and a half high.

CHAPTER IV. DISEASES MULES ARE LIABLE TO.--WHAT HE CAN DRAW, ETC., ETC.

The committee also say that the mule is a more steady animal in his draft than the horse. I think this the greatest mistake the committee has made. You have only to observe the manner in which a dray or heavily-loaded wagon will toss a mule about, and the way he will toss himself around on the road, to be satisfied that the committee have formed an erroneous opinion on that point. In starting with a load, the mule, in many cases, works with his feet as if they were set on a pivot, and hence does not take so firm a hold of the ground as the horse does. I have never yet seen a mule in a dray or cart that could keep it from jolting him round. In the first place, he has not the power to steady a dray; and, in the second place, they never can be taught to do it. In fine, they have not the formation to handle a dray or cart. What, then, becomes of the idea that they are as steady in drays or teams as the horse.

The committee also say that mules are not subject to such ailments as horses--spavin, glanders, ringbone, and bots. If I had the committee here, I would show its members that every other mule in the quartermasters' department, over fifteen and a half hands high, is either spavined, ringboned, or ill some way injured by the above-named diseases. The mule may not be so liable to spavin as the horse, but he has ringbone just the same. I cannot, for the life of me, see how the committee could have fallen into this error. There is this, however, to be taken into consideration: the mule is not of so sensitive a nature as the horse, and will bear pain without showing it in lameness. The close observer, however, can easily detect it. One reason why they do not show spavin and ringbone so much at the horse, is because our blacksmiths do not cut their heels as low as they do a horse's, and consequently that part of the foot is not made to work so hard. If you believe a mule has a ringbone, and yet is not lame, just cut his heel down low, and give him a few good pulls in a muddy place, and he will soon develop to you both lameness and ringbone. Cut his toes down and leave his heels high, and he will not be apt to go lame with it.

The committee also say that a Mr. Elliott, of the Patuxent Furnaces, says they hardly ever had a mule die of disease. This is a strange statement; for the poorest teams I ever saw, and the very worst bred stock, were on the Patuxent River, through the southern part of Maryland, and at the markets on Washington City. It is pitiable to see, as you can on market days, the shabby teams driven by the farmers of eastern and southern Maryland. A more broken-hearted, poverty-stricken, and dejected-looking set of teams can be seen nowhere else. The people of Maryland have raised good horses; it is high time they waked up to the necessity, and even profit, of raising a better kind of mule.

In regard to the draft power of mules, in comparison with horses, there are various opinions; and yet it is one which ought to be easily settled. I have tested mules to the very utmost of their strength, and it was very rare to find a pair that could draw thirty hundred weight a single year, without being used up completely. Now, it is well known that in the northern and western States you can find any number of pairs of horses that will draw thirty-five and forty hundred weight anywhere. And they will keep doing it, day after day, and retain their condition.

There was one great difficulty the Agricultural Committee of South Carolina had to contend with, and it was this. At the time it had the subject of the mule under consideration, he was not used generally throughout the United States. I can easily understand, therefore, that the committee obtained its knowledge from the very few persons who had them, and made the best report it could under the circumstances. Indeed, I firmly believe the report was written with the intention of giving correct information, but it failed entirely. In recommending any thing of this kind, great care should be taken not to lead the inexperienced astray, and to give only such facts as are obtained from thorough knowledge; and no man should be accepted as authority in the care and treatment of animals, unless he has had long experience with them, and has made them a subject of study.

A few words more on breaking the mule. Don't fight or abuse him. After you have harnessed him, and he proves to be refractory, keep your own temper, slack your reins, push him round, backward and forward, not roughly; and if he will not go, and do what you want, tie him to a post and let him stand there a day or so without food or water. Take care, also, that he does not lie down, and be careful to have a person to guard him, so that he does not foul in the harness. If he will not go, after a day or two of this sort of treatment, give him one or two more of it, and my word for it, he will come to his senses and do any thing you want from that time forward. Some persons assert that the mule is a very cunning animal; others assert that he is dull and stupid, and cannot be made to understand what you want. He is, I admit, what may be called a tricky animal; but, for experiment sake, just play one or two tricks with him, and he will show you by his action that he understands them well. Indeed, he knows a great deal more than he generally gets credit for, and few animals are more capable of appreciating proper treatment. Like many other species of animal, there are scarcely two to be found of precisely the same temper and disposition, if we except the single vice of kicking, which they will all do, especially when well fed and rested. And we can excuse even this vice in consideration of the fact, that the mule is not a natural animal, but only an invention of man. Some persons are inclined to think that, when a mule is a kicker, he has not been properly broken. I doubt if you can break a mule so that he will not kick a stranger at sight, especially if he be under six years old. The only way to keep a mule from kicking you is to handle it a great deal when young, and accustom it to the ways and actions of men. You must through kindness convince it that you are not going to harm or abuse it; and you can do that best by taking hold of it in a gentle manner every time it appears to be frightened. Such treatment I have always found more effective than all the beating and abusing you can apply.

There is another fault the mule has to contend against. It is the common belief among teamsters and others that he has less confidence in man than the horse has, and to improve this they almost invariably apply the whip. The reason for this want of confidence is readily found in the fact that mule colts are never handled with that degree of kindness and care that horse colts are. They are naturally more stubborn than the horse, and most of those persons who undertake to halter or harness them for the first time are even more stubborn in their disposition than the mule. They commence to break the animal by beating him in the most unmerciful manner, and that at once so excites the mule's stubbornness, that many of them, in this condition, would not move an inch if you were to cut them to pieces. And let me say here that nothing should be so much avoided in breaking this animal as the whip. The young, unbroken mule cannot be made to understand what you are whipping him for.

It is a habit with mule drivers in the army, many of whom are men without feeling for a dumb animal, to whip mules just to hear their whips crack, and to let others hear with what dexterity they can do it. It has a very bad effect on the animals, and some means should be applied to stop it. Army teamsters and stable-men seem to regard it as a virtue to be cruel to animals. They soon cultivate vicious habits, and a bad temper seems to grow up with their occupation. It naturally follows, then, that in the treatment of their animals they do just what they ought not to do. The Government has been a very severe sufferer by this; and I contend that during a war it is just as necessary to have experienced and well trained teamsters as it is to have hardened and well trained soldiers.

The mule is peculiar in his dislikes. Many of them, when first harnessed, so dislike a blind bridle that they will not work in it. When you find this, let him stand for say a day in the blinders, and then take them off, and in forty-nine cases out of fifty he will go at once.

It has been said that the mule never scares or runs away. This is not true. He is not so apt to get frightened and run away as the horse is. But any one who has had long experience with them in the army knows that they will both get frightened and run away. They do not, however, lose all their senses when they get frightened and run away, as the horse does. Bring a mule back after he has run away, and in most cases he will not want to do it again. A horse that has once run away, however, is never safe afterward. Indeed, in all the tens of thousands of mules that I have handled, I never yet found an habitual runaway. Their sluggish nature does not incline them to such tricks. If a team attempts to run away, one or two of them will fall down before they have gone far, and this will stop the remainder. Attempt to put one up to the same speed you would a horse, over a rough road, and you will have performed wonders if he does not fall and break your bones.

The mule, especially if large, cannot stand hard roads and pavements. His limbs are too small for his body, and they generally give out. You will notice that all good judges of road and trotting horses like to see a good strong bone in the leg. This is actually necessary. The mule, you will notice, is very deficient in leg, and generally have poor muscle. And many of them are what is called cat-hammed.

Working Condition of Mules.--Most persons, when they see a good, fat, slick mule, are apt to exclaim: "What a fine mule there is!" He takes it for granted that because the animal is fat, tall, and heavy, he must be a good work animal. This, however, is no criterion to judge by. A mule, to be in good condition for work, should never be any fatter than what is known as good working condition. One of fourteen and a half hands high, to be in good working condition, should not weigh over nine hundred and fifty pounds. One of fifteen hands high should not weigh over one thousand pounds. If he does, his legs will in a very short time give out, and he will have to go to the hospital. In working a mule with too much flesh, it will produce curbs, spavin, ringbone, or crooked hocks. The muscles and tendons of their small legs are not capable of carrying a heavy weight of body for any length of time. He may not, as I have said before, show his blemishes in lameness, but it is only because he lacks that fine feeling common to the horse. I have, singular as it may seem, known mules that have been spavined, curbed, and ringboned, and yet have been worked for years without exhibiting lameness.

Avoid spotted, or dapple mules; they are the very poorest animal you can get. They cannot stand hard work, and once they get diseased and begin to lose strength, there is no saving them. The Mexicans call them pintos, or painted mules. We call them calico Arabians or Chickasaws. They have generally bad eyes, which get very sore during the heat and dust of summer, when many of them go blind. Many of the snow-white mules are of the same description, and about as useless. Mules with the white muzzle, or, as some term it, white-nore white, and with white rings round the eyes, are also of but little account as work mules. They can stand no hardship of any kind. Government, at least, should never purchase them. In purchasing mules, you must look well to the age, form, height, eyes, size of bone and muscle, and disposition; for these are of more importance than his color. Get these right and you will have a good animal.

If any gentleman wants to purchase a mule for the saddle, let him get one bred closer after the mare than the jack. They are more docile, handle easier, and are more tractable, and will do what you want with less trouble than the other. If possible, also, get mare mules; they are much more safe and trusty under the saddle, and less liable to get stubborn. They are also better than a horse mule for team purposes. In short, if I were purchasing mules for myself, I would give at least fifteen dollars more for mare mules than I would for horse. They are superior to the horse mule in every way. One reason is, that they possess all their natural faculties, while you deprive the horse of his by altering.

The most disagreeable and unmanageable, and I was going to say useless, animal in the world, is a stud mule. They are no benefit to anybody, and yet they are more troublesome than any other animal. They rarely ever get fat, and are always fretting; and it is next to impossible to keep them from breaking loose and getting at mares. Besides, they are exceedingly dangerous to have amongst horses. They will frequently fly at the horse, like a tiger, and bite, tear, and kick him to pieces. I have known them to shut their eyes, become furious, and dash over both man and beast to get at a mare. It is curious, also, that a white mare seems to have the greatest attractions for them. I have known a stud mule to take a fancy to a white mare, and it seemed impossible to keep him away from her. Mules of all kinds, however, seem to have a peculiar fancy for white mares and horses, and when this attachment is once formed, it is almost impossible to separate them. If you want to drive a herd of five hundred mules any distance, turn a white or gray mare in among them for two or three days, and they will become so attached to her that you may turn them out, and they will follow her anywhere. Just let a man lead the mare, and with two men mounted you can manage the whole herd almost as well as if they were in a team. Another way to lead mules is, to put a bell on the mare's neck. The mules will listen for that bell like a lot of school children, and will follow its tinkling, with the same instinct.

Another curious thing about the mule is this: You may hitch him up to-day for the first time, and he may become sullen and refuse to go a step for you. This may be very provoking, and perhaps excite your temper; but do not let it, for ten chances to one, if you take him out of the harness to-day and put him in again to-morrow, that he will go right off, and do any thing you want him. It is best always to get a young mule well used to the harness before you try to work him in a team. When you get him so that he is not afraid of the harness, you may consider your mule two-thirds broke.

I have seen it asserted that a team of mules was more easily handled than a team of horses. It is impossible that this can be so, for the reason that you never can make a mule as bridle-wise as a horse. To further prove that this cannot be so, let any reinsman put as many mules together as there are horses in the "band wagon" of a show, or circus, and see what he can do with them. There is not a driver living who can rein them with the same safety that he can a horse, and for the very reason, that whenever the mule finds that he has the advantage of you, he will keep it in spite of all you can do.

Mule Raising.--I never could understand why it was that almost every person, that raises stock, recommends big, ugly gollips of mares, for mule-breeding. The principle is certainly a wrong one, as a little study of nature must show. To produce a good, well-proportioned mule, you must have a good, compact, and serviceable mare. It is just as necessary as in the crossing of any other animal. It certainly is more profitable to raise good animals than poor ones; and you cannot raise good mules from bad mares, no matter what the jack is. You invariably see the bad mare in the flabby, long-legged mule.

It has been held by some of our officers, that the mule was a better animal for Government service, because he required less care and feed than the horse, and would go longer without water. This, again, is a grave mistake. The mule, if properly taken care of, requires nearly as much forage as the horse, and should be groomed and cared for just the same. I refer now to team animals. Such statements do a great deal of injury, inasmuch as they encourage the men who have charge of animals to neglect and abuse them. The teamster who hears his superior talk in this way will soon take advantage of it. Animals of all kinds, in a wild and natural state, have a way of keeping themselves clean. If left wild, the mule would do it. But when man deprives them of the privileges by tying them up and domesticating them, he must assist them in the most natural way to keep themselves clean. And this assistance the animal appreciates to its fullest extent.

How to Handle a Mule Colt. --Owners and raisers of mules should pay more attention to their habits when young. And I would give them this advice: When the colt is six months old, put a halter on him and let the strap hang loose. Let your strap be about four feet long, so that it will drag on the ground. The animal will soon accustom himself to this; and when he has, take up the end and lead him to the place where you have been accustomed to feed him. This will make him familiar with you, and increase his confidence. Handle his ears at times, but don't squeeze them, for the ear is the most sensitive part of this animal. As soon as he lets you handle his ears familiarly, put a loose bridle on him. Put it on and take it off frequently. In this way you will secure the colt's confidence, and he will retain it until you need him for work.

Speaking of the sensitiveness of the mule's ear, a scratch, or the slightest injury to it, will excite their stubbornness and make them afraid of you. I have known a mule's ear to be scratched by rough handling, and for months afterward it was with the greatest difficulty you could bridle him. Nothing is more important than that you should bridle a young mule properly. I have found from experience that the best way is this: stand on the near side, of course; take the top of the bridle in your right hand, and the bit in your left; pass your arm gently over his eye until that part of the arm bends his ear down, then slip the bit into his mouth, and at the same time let your hand be working slowly with the bearings still on his head and neck, until you have arranged the head-stall.

It would be a saving of thousands of dollars to the Government, if, in purchasing mules, it could get them all halter and bridle-broken. Stablemen, in the employ of the Government, will not take the trouble to halter and bridle-break them properly; and I have seen hundreds of mules, in the City of Washington, totally ruined by tying them up behind wagons while young, and literally dragging them through the streets. These mules had never, perhaps, had a halter on before. I have seen them, while tied in this manner, jump back, throw themselves down, and be dragged on the ground until they were nearly dead. And what is worse, the teamster invariably seeks to remedy this by beating them. In most cases, the teamster would see them dragged to death before he would give them a helping hand. If he knew how to apply a proper remedy, very likely he would not give himself the trouble to apply it. I have never been able to find out how this pernicious habit of tying mules behind wagons originated; but the sooner an order is issued putting a stop to it, the better, for it is nothing less than a costly torture. The mule, more than any other animal, wants to see where he is going. He cannot do this at the tail of an army wagon, though it is an excellent plan for him to get his head bruised or his brains knocked out.

Some persons charge it as an habitual vice with the mule to pull back. I have seen horses contract that vice, and continue it until they killed themselves. But, in all my experience with the mule, I never saw one in which it was a settled vice. During the time I had charge of the receiving and issuing of horses to the army, I had a great many horses injured seriously by this vice of pulling back. Some of these horses became so badly injured in the spine that I had to send them to the hospital, then under the charge of Dr. L.H. Braley. Some were so badly injured that they died in fits; others were cured. Even when the mule gets his neck sore, he will endure it like the ox, and instead of pulling back, as the horse will, he will come right up for the purpose of easing it. They do not, as some suppose, do this because of their sore, but because they are not sensitive like the horse.

Packing Mules.--In looking over a copy of Mason's Farrier, or Stud Book, by Mr. Skinner, I find it stated that a mule is capable of packing six or eight hundred pounds. Mr. Skinner has evidently never packed mules, or he would not have made so erroneous a statement. I have been in all our Northern and Western Territories, in Old and New Mexico, where nearly all the business is done by pack animals, mules, and asses; and I have also been among the tribes of Indians bordering on the Mexican States, where they have to a great extent adopted the Spanish method of packing, and yet I never saw an instance when a mule could be packed six or eight hundred pounds. Indeed, the people in these countries would ridicule such an assertion. And here I purpose to give the result of my own experience in packing, together with that of several others who have long followed the business.

I also purpose to say something on what I consider the best mode of packing, the weight suitable for each animal, and the relative gain or loss that might result from this method of transportation, as compared with transportation by wagon. In the first place, packing ought never to be resorted to, because it cannot be done with profit, where the roads are good and wagons and animals are to be had. In mountains, over deserts and plains of sand, where forage is scant, and water only to be had at long intervals, then the pack is a necessity, and can be used with profit. Let it be understood, also, that in packing, the Spanish pack-mule, as as well as saddle, is the most suitable. Second: The Spanish method of packing is, above all others, the most ancient, the best and most economical. With it the animal can carry a heavier burden with less injury to himself. Third: The weight to be packed, under ever so favorable circumstances, should never be over four hundred and fifty pounds. Fourth: The American pack-saddle is a worthless thing, and should never be used when any considerable amount of weight is required to be packed.

If I had previously entertained any doubt in regard to this American pack-saddle, it was removed by what came under my observation three years ago. While employed in the quartermasters' depot, at Washington, D.C., as superintendent of the General Hospital Stables, we at one time received three hundred mules, on which the experiment of packing with this saddle had been tried in the Army of the Potomac. It was said this was one of General Butterfield's experiments. These animals presented no evidence of being packed more than once; but such was the terrible condition of their backs that the whole number required to be placed at once under medical treatment. Officers of the army who knew Dr. Braley, know how invariably successful he has been in the treatment of Government animals, and how carefully he treats them. Yet, in spite of all his skill, and with the best of shelter, fifteen of these animals died from mortification of their wounds and injuries of the spine. The remainder were a very long time in recovering, and when they did, their backs, in many cases, were scarred in such a manner as to render them unfit ever after for being used for a similar purpose. The use of the American pack-saddle, and lack of knowledge on the part of those in charge as to what mules were suitable for packing, did this. The experienced packer would have seen at a glance that a large portion of these mules were utterly unfit for the business. The experiment was a wretched failure, but cost the Government some thousands of dollars.

I ought to mention, however, that the class of mules on which this experiment was tried were loose, leggy animals, such as I have heretofore described as being almost unfit for any branch of Government service. But, by all means, let the Government abandon the American pack-saddle until some further improvements are made in it.

Now, as to the weight a mule can pack. I have seen the Delaware Indians, with all their effects packed on mules, going out on a buffalo hunt. I have seen the Potawatamies, the Kickapoos, the Pawnees, the Cheyennes, Pi-Ute, Sioux, Arapahoes, and indeed almost every tribe that use mules, pack them to the very extent of their strength, and never yet saw the mule that could pack what Mr. Skinner asserts. More than that, I assert here that you cannot find a mule that will pack even four hundred pounds, and keep his condition sixty days. Eight hundred pounds, Mr. Skinner, is a trying weight for a horse to drag any distance. What, then, must we think of it on the back of a mule? The officers of our quartermasters' department, who have been out on the plains, understand this matter perfectly. Any of these gentlemen will tell you that there is not a pack train of fifty mules in existence, that can pack on an average for forty days, three hundred pounds to the animal.

I will now give you the experience of some of the best mule packers in the country, in order to show that what has been written in regard to the mule's strength is calculated to mislead the reader. In 1856, William Anderson, a man whom I know well, packed from the City of Del Norte to Chihuahua and Durango, in Mexico, a distance of five hundred miles or thereabout. Anderson and a man of the name of Frank Roberts had charge of the pack train. They had seventy-five mules, and used to pack boxes of dry goods, bales, and even barrels. They had two Mexican drivers, and travelled about fifteen miles a day, at most, though they took the very best of care of their animals. Now, the very most it was possible for any mule in this train to get along with was two hundred and seventy-five pounds. More than this, they did not have over twenty-five mules out of the whole number that could pack two hundred and fifty pounds, the average weight to the whole train being a little less than two hundred pounds. To make this fifteen miles a day, they had to make two drives, letting the animals stop to feed whenever they had made seven or eight miles.

In 1858, this same Anderson packed for the expedition sent after the Snake Indians. His train consisted of some two hundred and fifty or three hundred mules. They packed from Cordelaine Mission to Walla Walla, in Oregon. The animals were of a very superior kind, selected for the purpose of packing out of a very large lot. Some of the very best of these mules were packed with three hundred pounds, but at the end of two weeks gave out completely.

In 1859, this same Anderson packed for a gentleman of the name of David Reese, living at the Dalles, in Portland, Oregon. His train consisted of fifty mules, in good average condition, many of them weighing nine hundred and fifty pounds, and from thirteen to fourteen hands high. His average packing was two hundred and fifty pounds. The distance was three hundred miles, and it occupied forty days in going and returning. Such was the severity of the labor that nearly two-thirds of the animals became poor, and their backs so sore as to be unfit for work. This trip was made from the Dalles, in Oregon, to Salmon Falls, on the Columbia River.

Anderson asserts it, as the result of his experience, that, in packing fifty mules a distance of three hundred miles with two hundred and fifty pounds, the animals will be so reduced at the end of the journey as to require at least four weeks to bring them into condition again. This also conforms with my own experience.

In 1857, there was started from Fort Laramie, Nebraska Territory, to go to Fort Bridger with salt, a train of forty mules. It was in the winter; each mule was packed with one hundred and eighty pounds, as near as we could possibly estimate, and the train was given in charge of a man of the name of Donovan. The weather and roads were bad, and the pack proved entirely too heavy. Donovan did all he could to get his train through, but was forced to leave more than two-thirds of it on the way. At that season of the year, when grass is poor and the weather bad, one hundred and forty or one hundred and fifty pounds is enough for any mule to pack.

There were also, in 1857, regular pack trains run from Red Bluffs, on the Sacramento River, in California, to Yreka and Curran River. Out of all the mules used in these trains, none were packed with over two hundred pounds. To sum up, packing never should be resorted to when there is any other means of transportation open. It is, beyond doubt, the most expensive means of transportation, even when the most experienced packers are employed. If, however, it were necessary for the Government to establish a system of packing, it would be a great saving to import Mexicans, accustomed to the work, to perform the labor, and Americans to take charge of the trains. Packing is a very laborious business, and very few Americans either care about doing it, or have the patience necessary to it.

CHAPTER V. PHYSICAL CONSTRUCTION OF THE MULE.

I now propose to say something on the mule's limbs and feet. It will be observed that the mule has a jack's leg from the knee down, and in this part of the leg he is weak; and with these he frequently has to carry a horse's body. It stands to reason, then, that if you feed him until he gets two or three hundred pounds of extra flesh on him, as many persons do, he will break down for want of leg-strength. Indeed, the mule is weakest where the horse is strongest. His feet, too, are a singular formation, differing very materially from those of the horse. The mule's feet grow very slow, and the grain or pores of the hoof are much closer and harder than those of the horse. It is not so liable, however, to break or crumble. And yet they are not so well adapted for work on macadamized or stony roads, and the more flesh you put on his body, after a reasonable weight, the more you add to the means of his destruction.

Observe, for instance, a farmer's mule, or a poor man's mule working in the city. These persons, with rare exceptions, feed their mules very little grain, and they are generally in low flesh. And yet they last a very long time, notwithstanding the rough treatment they get. When you feed a mule, you must adjust the proportions of his body to the strength of his limbs and the kind of service he is required to perform. Experience has taught me, that the less you feed a mule below what he will eat clean, just that amount of value and life is kept out of him.

In relation to feeding animals. Some persons boast of having horses and mules that eat but little, and are therefore easily kept. Now, when I want to get a horse or a mule, these small eaters are the last ones I would think of purchasing. In nine cases out of ten, you will find such animals out of condition. When I find animals in the Government's possession, that cannot eat the amount necessary to sustain them and give them proper strength, I invariably throw them out, to be nursed until they will eat their rations. Animals, to be kept in good condition, and fit for proper service, should eat their ten and twelve quarts of grain per head per day, with hay in proportion--say, twelve pounds.

I wish here again to correct a popular error, that the mule does not eat, and requires much less food than the horse. My experience has been, that a mule, twelve hands high, and weighing eight hundred pounds, will eat and, indeed, requires just as much as a horse of similar dimensions. Give them similar work, keep then in a stable, or camp them out during the winter months, and the mule will eat more than the horse will or can. A mule, however, will eat almost any thing rather than starve. Straw, pine boards, the bark of trees, grain sacks, pieces of old leather, do not come amiss with him when he is hungry. There were many instances, during the late war, where a team of mules were found, of a morning, standing over the remains of what had, the evening before, been a Government wagon. When two or more have been kept tied to a wagon, they have been known to eat each other's tail off to the bone, And yet the animal, thus deprived of his caudal appendage, did not evince much pain.

In the South, many of the plantations are worked with mules, driven by negroes. The mule seems to understand and appreciate the negro; and the negro has a sort of fellow-feeling for the mule. Both are sluggish and stubborn, and yet they get along well together. The mule, too, is well suited to plantation labor, and will outlast a horse at it. The soil is also light and sandy, and better suited to the mule's feet. A negro has not much sympathy for a work-horse, and in a short time will ruin him with abuse, whereas he will share his corn with the mule. Nor does the working of the soil on southern plantations overtax the power of the mule.

The Value of Harnessing properly.--In working any animal, and more especially the mule, it is both humane and economical to have him harnessed properly, Unless he be, the animal cannot perform the labor he is capable of with ease and comfort, And you cannot watch too closely to see that every thing works in its right place. Begin with the bridle, and see that it does not chafe or cut him, The army blind-bridle, with the bit alteration attached, is the very best bridle that can be used on either horse or mule. Be careful, however, that the crown-piece is not attached too tight. Be careful, also, that it does not draw the sides of the animal's mouth up into wrinkles, for the bit, working against these, is sure to make the animal's mouth sore. The mule's mouth is a very difficult part to heal, and once it gets sore he becomes unfit for work. Your bridle should be fitted well to the mule's head before you attempt to work him in it. Leave your bearing-line slack, so as to allow the mule the privilege of learning to walk easy with harness on. It is too frequently the case, that the eyes of mules that are worked in the Government's service are injured by the blinds being allowed to work too close to the eyes. This is caused by the blind-stay being too tight, or perhaps not split far enough up between the eyes and ears. This stay should always be split high enough up to allow the blinds to stand at least one inch and a half from the eye.

Another, and even more essential part of the harness is the collar. More mules are maimed and even ruined altogether by improperly fitting collars, than is generally believed by quartermasters. It requires more judgment to fit a collar properly on a mule than it does to fit any other part of the harness. Get your collar long enough to buckle the strap close up to the last hole. Then examine the bottom, and see that there be room enough between the mule's neck or wind-pipe to lay your open hand in easily. This will leave a space between the collar and the mule's neck of nearly two inches. Aside from the creased neck, mules' necks are nearly all alike in shape, They indeed vary as little in neck as they do in feet; and what I say on the collar will apply to them all, The teamster has always the means in his own hands of remedying a bad fitting collar. If the animal does not work easy in it, if it pinch him somewhere, let it remain in water over night, put it on the animal wet the next morning, and in a few minutes it will take the exact formation of the animal's neck. See that it is properly fitted above and below to the hames, then the impression which the collar takes in a natural form will be superior to the best mechanical skill of the best harness-maker.

There is another thing about collars, which, in my opinion, is very important. When you are pursuing a journey with teams of mules, where hay and grain are scarce, the animals will naturally become poor, and their necks get thin and small. If once the collar becomes too large, and you have no way of exchanging it for a smaller one, of course you must do the next best thing you can. Now, first take the collar off the animal, lay it on a level, and cut about one inch out of the centre. When you have done this, try it on the animal again; and if it still continues too large take a little more from each side of the centre until you get it right. In this way you can effect the remedy you need.

In performing a long journey, the animals will, if driven hard, soon show you where the collar ought to be cut, They generally get sore on the outer part of the shoulder, and this on account of the muscle wasting away. Teamsters on the plains and in the Western Territories cut all the collars when starting on a trip. It takes less time afterward to fit them to the teams, and to harness and unharness.

When you find out where the collar has injured the shoulder, cut it and take out enough of the stuffing to prevent the leather from touching the sore. In this way the animal will soon get sound-shouldered again. Let the part of the leather you cut hang loose, so that when you take the stuffing out you may put it back and prevent any more than is actually necessary from coming out.

See that your hames fit well, for they are a matter of great importance in a mule's drawing. Unless your hames fit your collar well, you are sure to have trouble with your harness, and your mule will work badly. Some persons think, because a mule can be accustomed to work with almost any thing for a harness, that money is saved in letting him do it. This is a great mistake. You serve the best economy when you harness him well and make his working comfortable. Indeed, a mule can do more work with a bad-fitting collar and harness than a man can walk with a bad-fitting boot. Try your hames on, and draw them tight enough at the top of the mule's neck, so that they will not work or roll round. They should be tight enough to fit well without pinching the neck or shoulder, and in fine, fit as neatly as a man's shirt-collar.

Do not get the bulge part of your collar down too low. If you do, you interfere with the machinery that propels the mule's fore legs. Again, if you raise it too high, you at once interfere with his wind. There is an exact place for the bulge of the collar, and it is on the point of the mule's shoulder. Some persons use a pad made of sheepskin on the toe of the collar. Take it off, for it does no good, and get a piece of thick leather, free from wrinkles, ten or twelve inches long and seven wide; slit it crosswise an inch or so from each end, leaving about an inch in the centre. Fit this in, in place of the pad of sheepskin, and you will have a cheaper, more durable, and cooler neck-gear for the animal. You cannot keep a mule's neck in good condition with heating and quilted pads. The same is true of padded saddles. I have perhaps ridden as much as any other man in the service, of my age, and yet I never could keep a horse's back in good condition with a padded saddle when I rode over twenty-five or thirty miles a day.

There is another evil which ought to be remedied. I refer now to the throat-latch. Hundreds of mules are in a measure ruined by allowing the throat-latch to be worked too tight. A tight throat-latch invariably makes his head sore. Besides, it interferes with a part which, if it were not for, you would not have the mule--his wind. I have frequently known mules' heads so injured by the throat-latch that they would not allow you to bridle them, or indeed touch their heads. And to bridle a mule with a sore head requires a little more patience than nature generally supplies man with.

Let a mule's ears alone. It is very common with teamsters and others, when they want to harness mules, to catch them by the ears, put twitches on their ears. Even blacksmiths, who certainly ought to know better, are in the habit of putting tongs and twitches in their ears when they shoe them. Now, against all these barbarous and inhuman practices, I here, in the name of humanity, enter my protest. The animal becomes almost worthless by the injuries caused by such practices. There are extreme cases in which the twitch may be resorted to, but it should in all cases be applied to the nose, and only then when all milder means have failed.

But there is another, and much better, method of handling and overcoming the vices of refractory mules. I refer to the lariat. Throw the noose over the head of the unruly mule, then draw him carefully up to a wagon, as if for the purpose of bridling him. In case he is extremely hard to bridle, or vicious, throw an additional lariat or rope over his head, fixing it precisely as represented in the drawing. By this method you can hold any mule. But even this method had better be avoided unless where it is absolutely necessary.

It is now August, 1866. We are working five hundred and fifty-eight animals, from six o'clock in the morning until seven o'clock at night, and out of this number we have not got ten sore or galled animals. The reason is, because we do not use a single padded saddle or collar. Also, that the part of the harness that the heaviest strain comes on is kept as smooth and pliable as it is possible for it to be. Look well to your drawing-chains, too, and see that they are kept of an even length. If your collar gets gummy or dirty, don't scrape it with a knife; wash it, and preserve the smooth surface. Your breeching, or wheel harness, is also another very important part; see that it does not cut and chafe the animal so as to wear the hair off, or injure the skin. If you get this too tight, it is impossible for the animal to stretch out and walk free. Besides obstructing the animal's gait, however, the straps will hold the collar and hames so tight to his shoulder as to make him sore on the top of his neck. These straps should always be slack enough to allow the mule perfect freedom when at his best walk.

And now I have a few words to say on Government wagons. Government wagons, as now made, can be used for other purposes besides the army. The large-sized Government wagon is, it has been proved, too heavy for four horses. The smaller sized one is nearer right; but whenever you take an ordinary load on it (the smaller one) and have a rough country to move through, it will give out. It is too heavy for two horses and a light load, and yet not heavy enough to carry twenty-five hundred or three thousand pounds, a four-horse load, when the roads are in any way bad. They do tolerably well about cities, established posts, and indeed anywhere where the roads are good, and they are not subject to much strain. Improvements on the Government wagon have been attempted, but the result has been failure. The more simple you can get such wagons, the better, and this is why the original yet stands as the best. There is, however, great difference in the material used, and some makers make better wagons than others. The six and eight-mule wagon, the largest size used for road and field purposes, is, in my humble opinion, the very best adapted to the uses of our American army. During the rebellion there were a great many wagons used that were not of the army pattern. One of these, I remember, was called the Wheeling wagon, and used to a great extent for light work, and did well. On this account many persons recommended them. I could not, and for this reason: they are too complicated, and they are much too light to carry the ordinary load of a six-mule team. At the end of the war it was shown that the army pattern wagon had been worked more, had been repaired less, and was in better condition than any other wagon used. I refer now to those made in Philadelphia, by Wilson & Childs, or Wilson, Childs & Co. They are known in the army as the Wilson wagon. The very best place to test the durability of a wagon is on the plains. Run it there, one summer, when there is but little wet weather, where there are all kinds of roads to travel on and loads to carry, and if it stands that it will stand any thing. The wagon-brake, instead of the lock-chain, is a great and very valuable improvement made during the War. Having a brake on the wagon saves the time and trouble of stopping at the top of every hill to lock the wheels, and again at the bottom to unlock them. Officers of the army know how much trouble this used to cause, how it used to block up the roads, and delay the movements of troops impatient to get ahead. The lock-chain ground out the wagon tire in one spot. The brake saves that; and it also saves the animal's neck from that bruising and chafing incident to the dead strain that was required when dragging the locked wheel.

There is another difficulty that has been overcome by the wagon-brake. In stopping to lock wheels on the top of a hill, your train get into disorder. In most cases, when trains are moving on the road, there is a space of ten or fifteen feet between the wagons. Each team, then, will naturally close up that space as it comes to the place for halting to lock. Now, about the time the first teamster gets his wheel locked, the one in the rear of him is dismounting for the same purpose. This being repeated along the train, it is not difficult to see how the space must increase, and irregularity follow. The more wagons you have to lock with the drag-chain, the further you get the teams apart. When you have a large body of wagons moving together, it naturally follows that, with such a halt as this, the teams in the rear must make twenty-five halts, or stops, and starts, for everyone that the head team makes.

When the teamster driving the second team gets ready to lock, the first, or head team, starts up. This excites the mule of the second to do the same, and so all along the train. This irritates the teamster, and he is compelled to run up and catch the wheel-mules by the head, to make them stop, so that he can lock his wheels. In nine cases out of ten he will waste time in punishing his animals for what they do not understand. He never thinks for a moment that the mule is accustomed to start up when the wagon ahead of him moves, and supposes he is doing his duty. In many cases, when he had got his wheels locked, he had so excited his mules that they would run down the hill, cripple some of the men, break the wagon, cause a "smash-up" in the train, and perhaps destroy the very rations and clothes on which some poor soldier's life depended. We all know what delay and disaster have resulted from the roads being blocked up in this manner. The brake, thanks to the inventor, offers a remedy for all this. It also saves the neck and shoulders of every animal in the train; it saves the feet of the wheelers; it saves the harness; it saves the lead and swing mules from being stopped so quick that they cut themselves; and it saves the wheels at least twenty per cent. Those who have had wagons thrown over precipices, or labored and struggled in mud and water two and three hours at a time, can easily understand how time and trouble could have been saved if the wagon could have been locked in any way after it started over those places. The best brake by all odds, is that which fastens with a lever chain to the brake-bar. I do not like those which attach with a rope, and for the reason that the lazy teamster can sit on the saddle-mule and lock and unlock, while, with the chain and lever, he must get off. In this way he relieves the saddle-mule's back.

We all know that, in riding mules down steep or long hills, you do much to stiffen them up and wear them out.

CHAPTER VI. SOMETHING MORE ABOUT BREEDING MULES.

Before I close this work, I desire to say something more about breeding mules. It has long been a popular error that to get a good mule colt you must breed from large mares. The average sized, compact mare, is by all odds the superior animal to breed mules from. Experience has satisfied me that very large mules are about as useless for army service as very large men are for troopers. You can get no great amount of service out of either. One is good at destroying rations; the other at lowering haystacks and corn-bins. Of all the number we had in the army, I never saw six of these large, overgrown mules that were of much service. Indeed, I have yet to see the value in any animal that runs or rushes to an overgrowth. The same is true with man, beast, or vegetable. I will get the average size of either of them, and you will acknowledge the superiority.

The only advantage these large mares may give to the mule is in the size of the feet and bone that they may impart. The heavier you can get the bone and feet, the better. And yet you can rarely get even this, and for the reason that I have before given, that the mare, in nineteen cases out of twenty, breeds close after the jack, more especially in the feet and legs. It makes little difference how you cross mares and jacks, the result is almost certain to be a horse's body, a jack's legs and feet, a jack's ears, and, in most cases, a jack's marks.

Nature has directed this crossing for the best, since the closer the mare breeds after the jack the better the mule. The highest marked mules, and the deepest of the different colors, I have invariably found to be the best. What is it, let us inquire, that makes the Mexican mule hardy, trim, robust, well-marked after the jack, and so serviceable? It is nothing more nor less than breeding from sound, serviceable, compact, and spirited Mexican or mustang mares. You must, in fact, use the same judgment in crossing these animals as you would if you wanted to produce a good race or trotting horse.

We are told, in Mason and Skinner's Stud Book, that in breeding mules the mares should be large barrelled small limbed, with a moderate-sized head and a good forehead. This, it seems to me, will strike our officers as a very novel recommendation. The mule's limbs and feet are the identical parts you want as large as possible, as everyone that has had much to do with the animal knows. You rarely find a mule that has legs as large as a horse. But the mule, from having a horse's body, will fatten and fill up, and become just as heavy as the body of an average-sized horse. Having, then, to carry this extra amount of fat and flesh on the slender legs and feet of a jackass, you can easily see what the result must be. No; you will be perfectly safe in getting your mule as large-legged as you can. And by all means let the mare you breed from have a good, sound, healthy block of a foot. Then the colt will stand some chance of inheriting a portion of it. It is natural that the larger you get his feet the steadier he will travel. Some persons will tell you that these small feet are natural, and are best adapted to the animal. But they forget that the mule is not a natural animal, only an invention of man. Let your mare and jack be each of the average size, the jack well marked, and No. 1 of his kind, and I will take the product and wear out any other style of breed. Indeed, you have only to appeal to your better judgment to convince you as to what would result from putting a jack, seven or eight hands high, to a mare of sixteen or more.

I have witnessed some curious results in mule breeding, and which it may be well enough to mention here. I have seen frequent instances where one of the very best jacks in the country had been put to mares of good quality and spirit. Putting them to such contemptible animals seemed to degrade them, to destroy their natural will and temper. The result was a sort of bastard mule, a small-legged, small-footed, cowardly animal, inheriting all the vices of the mule and none of the horse's virtues-- the very meanest of his kind.

CHAPTER VII. ANCIENT HISTORY OF THE MULE.

The mule seems to have been used by the ancients in a great variety of ways; but what should have prompted his production must for ever remain a mystery. That they early discovered his great usefulness in making long journeys, climbing mountains, and crossing deserts of burnings and, when subsistence and water were scarce, and horses would have perished, is well established. That he would soon recover from the severe effects of these long and trying journeys must also have been of great value in their eyes. But however much they valued him for his usefulness, they seem not to have had the slightest veneration for him, as they had for some other animals. I am led to believe, then, that it was his great usefulness in crossing the sandy deserts that led to his production. It is a proof, also, that where the ass was at hand there also was the horse, or the mule could not have been produced. Any people with sufficient knowledge to produce the mule would also have had sufficient knowledge to discover the difference between him and the horse, and would have given the preference to the horse in all service except that I have just described. And yet, in the early history of the world, we find men of rank, and even rulers, using them on state and similar occasions; and this when it might have been supposed that the horse, being the nobler animal, would have made more display.

The Scriptures tell us that Absalom, when he led the rebel hosts against his father David, rode on a mule, that he rode under an oak, and hung himself by the hair of his head. Then, again, we hear of the mule at the inauguration of King Solomon. It is but reasonable to suppose that the horse would have been used on that great occasion, had he been present. On the other hand, it is not reasonable to suppose that the ass, or any thing pertaining to him, was held in high esteem by a nation that believed they were commanded by God, through their prophet Moses, not to work the ox and the ass together. It must be inferred from this that the ass was not held in very high esteem, and that the prohibition was for the purpose of not degrading the ox, he being of that family of which the perfect males were used for sacrifice. The ass, of course, was never allowed to appear on the sacred altar. And yet He who came to save our fallen race, and open the gates of heaven, and fulfil the words of the prophet, rode a female of this apparently degraded race of animals when He made his triumphal march into the city of the temple of the living God.

List of Mules Received, died, and Shot, at the Depot of Washington, D.C., from 1st February, 1863. to 31st July, 1866.

Month	1863			1864			1865		
	Received	Died	Shot	Received	Died	Shot	Received	Died	Shot
Jan.	..	..	..	624	14	76	3,677	66	226
Feb.	135	96	7	329	16	62	1,603	84	150
Mar.	2,552	150	4	448	10	64	2,823	77	169
Apr.	2,906	118	61	1,305	15	47	6,102	106	223
May.	1,087	56	46	2,440	18	52	11,780	68	211
Jun.	3,848	120	118	4,410	76	48	19,304	178	49
Jul.	1,731	94	335	4,702	74	125	13,398	462	68
Aug.	5,250	51	159	5,431	88	231	1,275	284	23
Sep.	2,834	72	248	1,198	64	176	1,536	3	18
Oct.	1,166	36	202	1,468	81	134	876	..	..
Nov.	2,934	30	204	3,036	35	123	252	3	..
Dec.	2,832	14	113	3,923	66	158	324	4	..
Total	27,275	837	1,497	29,414	557	1,296	62,950	1,335	1,137

1866
Received Died Shot

169	..	..
34	2	1
13	..	..
29	1	..
20	1	..
2	..	..
62	..	..
..	..	..
..	..	..
..	..	..
..	..	..
..	..	..
329	4	1

DATE	RECEIVED	DIED	SHOT
1863.............	27,275	837	1,497
1864.............	29,414	557	1,296
1865.............	62,950	1,335	1,137
1866.............	329	4	1
Total...........	119,968	2,733	3,931

PICTURES OF SOME OF OUR MOST CELEBRATED ARMY MULES.

I have had photographs taken of some of our mules. A number of these animals performed extraordinary service in connection with the Army of the Potomac and the Western Army. One of them, a remarkable animal, made the great circuit of

Sherman's campaign, and has an historical interest. I propose to give you these illustrations according to their numbers.

No.1, then, is a very remarkable six-mule team. It was fitted out at Berryville,
Maryland, early in the spring of 1861, under the directions of Captain Sawtelle, A.
Q. M. They are all small, compact mules, and I had them photographed in order to
show them together. The leaders and swing, or, as some call them, the middle leaders, have been worked steadily together in the same team since December 31, 1861.
They have also been driven by the same driver, a colored man, of the name of Edward Wesley Williams. He was with Captain Sawtelle until the 1st of March, 1862;
was then transferred, with his team, to the City of Washington, and placed under a
wagon-master of the name of Horn, who belonged to Harrisburg, Pa. Wesley took
good care of his team, and was kept at constant work with it in Washington, until
May 14, 1862. He was then transferred, with his team, to a train that was ordered
to join General McClellan at Fort Monroe. He then followed the fortunes of the
Army of the Potomac up the Peninsula; was at the siege of Yorktown, the battle of
Williamsburg, and in the swamps of the Chickahominy. He was also in the seven
days' battles, and brought up at Harrison's Landing with the Army of the Potomac.
He then drove his team back to Fort Monroe, where they were shipped, with the
animals of the Army of the Potomac, for Washington. He was set to work as soon as
he reached a landing, and participated in hauling ammunition at the second battle
of Bull Run. He then followed the army to Antietam, and from that battle-field to
Fredericksburg, where he hauled ammunition during the terrible disaster under
General Burnside. The team then belonged to a train of which John Dorny was
wagon-master. When General Hooker took command of the army this team followed him through the Chancellorville and Chantilly fights. It also followed the
Army of the Potomac until General Grant took command, when the train it belonged to was sent to City Point. This brings us up to 1864. It was with the army in
front of Petersburg, and, during that winter, the saddle mule was killed by the en-
emy's shot while the team was going for a load of wood. In short, they were worked
every day until Richmond was taken. In June, 1865, they were transferred back to
the City of Washington. It is now August, 1866, and they are still working in the
train, and make one of the very best teams we have. I refer now to the leaders and
swing mules, as they are the only four that are together, and that followed the

1

2

Army of the Potomac through all its campaigns. There is not a mule of the four that is over fourteen and a half hands high, and not one that weighs over nine hundred pounds. This team, I ought to add here, has frequently been without a bite of hay or grain for four or five days, and nothing to eat but what they could pick up along the road. And there are instances when they have been twenty-four hours without a sup of water. The experienced eye will see that they have round, compact bodies, and stand well on their feet.

No. 2 is the leader of the team, and for light work on the prairies, packing, or any similar work, is a model mule. Indeed, she cannot be surpassed. Her bone and muscle is full, and she is not inclined to run to flesh.

No. 3 is the off-leader of the same team. She is a good eater, tough, hardy, and a good worker,--in every way a first-class mule. I would advise persons purchasing mules to notice her form. She is a little sprung in the knees; but this has in no way interfered with her working. This was occasioned by allowing the heels on her fore-feet to grow out too much. During, and for some time after, the second battle of Bull Run, the train to which she belonged was kept at very hard work. The shoes that were on her at that time, to use the driver's own language, were "put on to stay." Indeed, he informed me that they were on so long, that he concluded they had grown to the feet. And in this case, as in many others, for want of a little knowledge of the peculiarities of a mule's feet, and the injury that results from over-growth, the animal had to suffer, and was permanently injured.

No. 4 is the off-swing, or middle-leader mule. She is perfectly sound, of good height, a good eater, and a great worker. She is also well adapted for packing, and a tolerably good rider. Her ears and eyes are of the very finest kind, and her whole head indicates intelligence. Her front parts are perfection itself. She is also remarkably kind.

No. 5 is the near swing mule, or middle leader. She is what is called a mouse-color, and is the fattest mule in the team. She underwent the entire campaigns of the Army of the Potomac, and is to-day without a blemish, and capable of doing as much work as any mule in the pack. Her powers of endurance, as well as her ability to withstand starvation and abuse, are beyond description. I have had mules of her build with me in trains, in the Western Territories, that endured hardship and starvation to an extent almost incredible; and yet they were remarkably kind when

3

4

well treated, and would follow me like dogs, and, indeed, try to show me how much they could endure without flinching.

No. 6 is an off-wheel mule, of ordinary quality. I had to take the spotted mules from the wheels of this team, as they were not equal to the work required of them, and got very sore in front.

No. 7 is a spotted, or, as the. Mexicans call them, a calico mule. He and his mate were sent to the Army of the Potomac about the time General Grant took command of it. They were worked as wheel mules in the team until 1866, when this one, like nearly all spotted animals, showed his weak parts by letting up in his fore-feet, which became contracted to such an extent that the surgeon had to cut them nearly off. We were compelled to let him go barefoot until they grew out. This is one of the spotted mules I have referred to before. You never can rely on them.

No. 8 is the mate of No. 7. His bead, ears, and front shoulder indicate him to be of Canadian stock. His neck and front shoulder, as you will see, are faultless. But on looking closely at his eyes you will find them to be sore, and running water continually. I have noticed that nearly all animals in the army that are marked in this way have weak and inflamed eyes. A farmer should never purchase them.

No. 9 is a swing mule that has undergone a great deal of hardship. She is tolerably well formed but inclined to kick. She is also hard to keep in good condition, and unless great care is taken with her she would give out in the hind feet, where she now shows considerable fullness. When a mule's neck lacks the ordinary thickness there must be some direct cause for it, and you should set about finding out what it is. Lack of food is sometimes the cause. But in my opinion creased neck very frequently so affects the passages to and from the head, that the organs that should work in depositing flesh, fat, or muscle become deranged, and the neck becomes weak and in a disordered state. Purchasers would do well to discard these creased-neck mules.

No. 10 is an animal of an entirely different character from No. 9. She is remarkably gentle and tractable, of good form, and great endurance, and will work in any way. She is fifteen hands and one inch high, weighs ten hundred and fifty pounds, and is seven years old. This celebrated animal went through all of General Sherman's campaigns, and is as sound and active to-day as a four-year old.

No. 11 is one of those peculiar animals I have described elsewhere. He is all

5

6

bones and belly. His legs are long, and of little use as legs. He is five years old, sixteen and a half hands high, and weighs thirteen hundred and ninety pounds. One of his hind legs shows a thorough pin. His hocks are all out of shape, and his legs are stuck into his hoofs on nearly the same principle that you stick a post into the ground. The reason why his pastern-joints show so straight is, that the heels on the hind feet have been badly trimmed when shaving. They too have been permitted to grow too long, and thus he is thrown into the position you now see him. This mule belongs to a class that is raised to a considerable extent, and prized very highly in Pennsylvania. In the army they were of very little use except to devour forage.

No. 12 is what may be called a pack mule of the first class. He is seven years old, fifteen and a half hands high, and weighs eleven hundred and fifty-six pounds. This animal has endured almost incredible hardships. He is made for it, as you will readily see. He is what is called a portly mule, but is not inclined to run to belly unless over-fed and not worked. He has a remarkably kind disposition, is healthy, and a good feeder. This animal has but one evil to contend with. His off hind foot has grown too long, and plainly shows how much too far back it throws the pastern-joint. This is in a measure the effect of bad shoeing. It is very rare to find a blacksmith who discovers this fact until it is too late. Now there is nothing more easy than to ruin a mule by letting his toes grow too long. Doctor L.H. Braley, chief veterinary surgeon of the army, is now developing a plan for shoeing mules, which I consider the very best that has been suggested. His treatment of the foot when well, and how to keep it so; and how to treat the foot by shoeing when it becomes injured, is the best that can be adopted.

No. 13 is a mule that has been worked in a two-mule train which has been in my charge for about a year. She was previously worked in a six-mule train, as the off-wheel mule. She is five years old, rising; size, fifteen hands and three inches high, and weighs fourteen hundred and twenty-two pounds. She was received into the Government service at Wheeling, Virginia, and when shipped or transferred to this depot, with four hundred others, was but two years old, rising three. She was worked, at least a year or more, too young; and to this cause I attribute certain injuries which I shall speak of hereafter. This mule, with two hundred others, was transferred to the Army of the Potomac, and went through its campaigns from 1864 up to the fall of Richmond. She is an excellent worker, and her neck, head, and

7

8

fore shoulders are as fine as can be. Indeed, they are a perfect development of the horse. But her hips or flank joints are very deficient. Owing to her being worked too young, the muscles of the hind legs have given way, and they have become crooked. This is done frequently by the animal being placed as a wheeler when too young, and holding back under a heavy load. If you want to see how quick you can ruin young mules, place them in the wheels.

No. 14 is the off-wheel mule of a six-mule team. I had this mule photographed for the purpose of showing the effects of hitching animals so short to the team that the swingle-tree will strike or rest on their hocks. I referred to this great evil in another place. This mule is but six years old, sixteen hands high, and weighs nearly sixteen hundred pounds. Aside from the hocks, she is the best made and the best looking mule in the park; and is also a remarkably good worker. You will notice, however, that the caps of her hocks are so swollen and calloused by the action of the swingle-tree as to make them permanently disfigured. The position I have placed this mule in, as relates to the wagon wheel, is the proper position to put all wild, green, contrary or stubborn mules in when they are hard to bridle.

This is the severest use to which a lariat can be put on mule or horse. The person using it, however, should be careful to see that it sets well back to the shoulder of the animal. I refer now to the part of the loop that is around the neck. The end of the lariat should always be held by a man, and not made fast to any part of the wagon, so that if the animal falls or throws himself, you can slack up the lariat and save him from injury. Three applications of the buck will conquer them so thoroughly that you will have little trouble afterwards. Be careful to keep the lariat, in front, as high as the mule's breast; and see also that they are pulled up close to the front wheel before pulling it through the hind wheel.

DISEASES COMMON TO THE MULE, AND HOW THEY SHOULD BE TREATED.

The mule does not differ materially from the horse in the diseases he is afflicted with. He however suffers less from them, owing to lack of sensibility. It may be useful here to make a few remarks on the various diseases he is subject to, and to recommend a course of treatment which I have practiced and seen practiced, and

9

10

which I believe is the best that can be applied.

DISTEMPER IN COLTS.

This disease is peculiar to young mules. Its symptoms develop with soreness and swelling of the glands of the throat, a cough, difficulty of swallowing, discharging at the nostrils, and general prostration. If not properly treated it is surely fatal.

TREATMENT:--Give light bran mashes, plenty of common salt, and keep the animal in a warm and dry stable. You need not clothe, for the mule, unlike the horse, is not used to clothing. If the swelling under the throat shows a disposition to ulcerate, which it generally does, do nothing to prevent it. Encourage the ulcer, and let it come to a head gradually, for this is the easiest and most natural way that the trouble, which at first seems to pervade the whole system, can be got rid of. When the ulcer appears soft enough to lance, do so, and be careful to avoid the glands and veins. Lance through the skin in the soft spot, which appears almost ready to break. If the throat is at any time so swollen as to render swallowing difficult, give water frequently, about milk warm, with nourishing feed of oats, corn, or rye meal--the last is the best. If this treatment, which is very simple, be carefully carried out, few animals will fail to recover.

CATARRH OR COLDS.

This disease seldom attacks the mule. We have had many thousands of them in camp, and out of the whole number, I do not recollect of a case where it either destroyed or disabled a single animal. In fact, it is a question with me whether mules will take cold when kept as the Government keeps them--camped out, or standing in sheds where the temperature is the same as outdoors.

GLANDERS.

This is one of the most destructive of diseases with which the horse family is afflicted, and one that has set the best veterinary skill of the world at defiance. A remedy for it has yet to be discovered. I have deemed it proper here, however, to carefully describe its symptoms, and to recommend that all animals showing symptoms of it be kept by themselves until their case be definitely ascertained. When you have ascertained to a certainty that they are afflicted with the disease, destroy them as quick as possible. See, too, that the place where they have been kept is thoroughly cleansed and sprinkled with lime, for the disease is contagious and the slightest particle of virus will spread it anew. Farcy is but one stage of this terrible

11

12

disease, but is not necessarily fatal while in this stage. It should, however, be treated with great care and caution. Farcy can also be conveyed to others by inoculation. Any one who has had the field for observation the author has for the last four years, would become convinced that the recommendations I am about to make describe the only course to be taken with this contagious disease. The number of its victims under my observation were counted by thousands. All that can be done is to prevent, if possible, the disease taking place, and to destroy when ascertained to a certainty that the animal has contracted it. I would say here, however, that this subject will soon be thoroughly handled in a work soon to be published by Doctor Braley, head veterinary surgeon of the army. He will undoubtedly throw some light on the subject that has not yet appeared in print.

SYMPTOMS.

First:--When it appears in a natural form, without the agency of contagion or inoculation, dryness of the skin, entire omission of insensible perspiration, starring of the coat. Sometimes slight discoloring can be observed about the forehead and lower part of the ears. Drowsiness, want of lustre in the eye, slight swelling on the inside of the hind legs, extending up to the bu-boa. This condition of things may continue for several days, and will be followed by enlargement between the legs. The inflammation incident to this may entirely subside, or it may continue to enlarge, and break out in ulcers on the *lactiles* of the lymphatic, which accompanies the large veins. In the last case it has appeared in the form of Farcy. This being the case, the countenance assumes a more cheerful look, and the animal otherwise shows signs of relief from the discharges of poisonous matter. If it remain in this state, death is not generally the result. If the system be toned up it will sometimes heal, and the animal will seem to be in a recovering state of health. Yet, from watching the symptoms and general health of the animal afterwards, you will be convinced that the disease is only checked, not eradicated. Acting in the system, it only waits a favorable opportunity to act as a secondary agent in colds, general debility, or exposure, when it will make its appearance and produce death.

But in the first case, as shown by the swelling in the hind legs, if the swelling disappear, and general debility of the system continues; if the eyes grow more drowsy, and discharge from the lower corners; and if this is followed by discharge from the nostrils, slight swelling and hardening of the sub-maxillary glands, which

13

14

are between the under jaws, then it is clearly developed glanders. All the glands in the body have now become involved or poisoned, and death must follow in the course of ten or fifteen days, as the constitution of the animal may not be in a condition to combat the disease.

If this disease be annoyed by inoculation from the *farcy heads* of farcied animals into suppurating sores on other animals, it will be very slow in its progress, especially if it attack the other in a region remote from the lymphatic. If in a saddle-gall, it will make sores very difficult to heal. If there is any such thing as checking the disease in its progress, it is in these three cases.

I have observed that when it has been taken in a sore mouth it has followed down the cheek to the sub-maxillary gland, and ended in a clear case of glanders or farcy. There is another form in which this disease can be taken, and which is, of all others, the most treacherous and dangerous, yet never producing death without the agency of other diseases--always carrying with it the germs of infection, and ready to convey it to debilitated subjects and cause their death. The animal will still live himself, and show no sign of disease further than I am about to describe in the position. It is that which is taken in at the nostrils and attacks the sub-maxillary glands, which become enlarged and will remain so. When these become overloaded there will be a discharge at the nose. That being thrown off, it may be some time before any further discharge will be seen from the same source. In some cases, when the discharge is constant, this can be easily distinguished from gleet or ozena, from the healthy and natural appearance of the membranes of the nose, which at first are pale, then become fiery red or purple. In gleet the discharges from the nostrils, as in ozena, are of a very light color. In glanders they are first of a deep yellow, then of a dirty gray--almost slate color.

Mules affected with glanders of this kind, although it may seem hard from their otherwise healthy appearance, should be destroyed. They indeed carry with them the germs of infection and death, without any visible marks in their appearance to warn those who have the care of animals against their danger.

TEETHING.

As mules seldom change hands to any great extent until two or three years old, it is not deemed necessary here to say any thing of their age until they have reached two years, so as to give the inexperienced a wider scope. The mule's mouth undergoes exactly the same changes as the horse's. Between the ages of two and three these changes begin to take place in the mule's mouth. The front incisor teeth, two above and two below, are replaced by the horse for permanent teeth. These teeth are larger than the others, have two grooves in the outer converse surface, and the mark is long, narrow, deep, and black. Not having attained their full growth, they are somewhat lower than the others, the mark in the two next nippers being nearly worn out, and is also wearing away in the corner nippers.

A mule at three years old ought to have the central permanent nippers growing, the other two pairs uniting, six grinders in each jaw, above and below, the first and fifth level with the others, and the sixth protruding. As the permanent nippers wear and continue to grow, a narrow portion of the cone-shaped tooth is exposed to the attrition; and they look as if they had been compressed. This is not so, however; the mark of some gradually disappears as the pit is worn away. At the age of three and a half or four years the next pair of nippers will be changed, and the mouth at that time cannot be mistaken. The central nippers will have nearly attained their full growth, and a vacuity will be left where the second stood; or, they will begin to peep above the gum, and the corner ones will be diminished in breadth and worn down, the mark becoming small and faint. At this period also the second pair of grinders will be shed. At four years the central nippers will be fully developed, the sharp edges somewhat worn off, and the marks shorter, wider, and fainter. The next pair will be up, but they will be small, with the mark deep and extending quite across. Their corner nippers will be larger than the inside ones, yet smaller than they were, and flat, and nearly worn out. The sixth grinder will have risen to a level with the others; and the tushes will begin to appear in the male animal. The female seldom has them, although the germ is always present in the jaw. At four years and a half, or between that and five, the last important change takes place in the mouth of the mule. The corner nippers are shed, and the permanent ones begin to appear. When the central nippers are considerably worn, and the next pair are showing marks of wear, the tush will have protruded, and will generally be a full half inch in height. Externally it has a rounded prominence, with a groove on either side, and is evidently hollow within. At six years old the mark on the central nippers is worn out. There will, however, still be a difference of color in the center of the tooth. The cement filling up the hole made by the dipping in of the enamel, will present a browner hue than the other part of the tooth. It will be surrounded by an edge of enamel, and there will remain a little depression in the center, and also a depression around the case of the enamel. But the deep hole in the center of the enamel, with the blackened surface it presents, and the elevated edge of the enamel, will have disappeared. The mule may now be said to have a perfect mouth, all the teeth being produced and fully grown.

What I have said above must not be taken as a positive guide in all cases, for mules' mouths are frequently torn, twisted, smashed, and knocked into all kinds of shapes by cruel treatment, and the inexperience, to use no harsher term, of those who have charge of them. Indeed, I have known cases of cruelty so severe that it were impossible to tell the age of the animal from his teeth.

At seven years old the mark, in the way in which I have described it, is worn out in the four central nippers, and is also fast wearing away in the corner teeth. I refer now to a natural mouth that has not been subjected to injuries. At eight years old the mark is gone from all the bottom nippers, and may be said to be quite out of the mouth. There is nothing remaining in the bottom nippers by which the age of the mule can be positively ascertained. The tushes are a poor guide at any time in the life of the animal to ascertain his age by; they, more than any other of the teeth, being most exposed to the injuries I have referred to. From this time forward, the changes that take place in the teeth may be of some assistance in forming an opinion; but there are no marks in the teeth by which a year, more or less, can be positively ascertained. You can ascertain almost as much from the general appearance of the animal as from an examination of the mouth. The mule, if he be long-lived, has the same effect in changing his general appearance from youth to old age as is shown on the rest of the animal creation.

DISEASES OF THE TEETH.

There are few if any diseases to which the mule's teeth are subject, after the permanent teeth are developed; but during the time of their changes I have been led to believe that he suffers more inconvenience, or at least as much as any other animal--not so much on account of the suffering that nature inflicts upon him, as through the inexperience and cruelty of those who are generally intrusted with his care. I will here speak first of lampass. The animal's mouth is made sore and sensitive by teething; and this irritation and soreness is increased by the use of improper bits. As if this were not enough, resort is had to that barbarous and inhuman practice of burning out lampass. This I do, and always have protested against. If the gums are swollen from the cutting of teeth, which is about all the cause for their inflamed and enlarged appearance, a light stroke of a lancet or sharp knife over the gums, at a point where the teeth are forcing their way through, and a little regard to the animal's diet, will be all that is necessary. It must not be forgotten, that at this time the animal's mouth is too sore and sensitive to masticate hard food, such as corn. With the development of the teeth, however, the lampass will generally disappear.

THE EYE.

Mules are remarkable for having good eyes. Occasionally they become inflamed and sore. In such cases the application of cold water, and the removing of the cause, whether it be from chafing of the blinders, forcing the blood to the head through the influence of badly fitting collars, or any other cause known, is all I can recommend in their case.

THE TONGUE.

Mules suffer much from injury to the tongue, caused by the bad treatment of those who have charge of them, and also from sore month, produced in the same manner. The best thing for this is a light decoction of white-oak bark, applied with a sponge to the sore parts. Charcoal, mixed in water, and applied in the same manner, is good. Any quantity of this can be used, as it is not dangerous. If possible, give the animal nourishing gruels, or bran mashes; and, above all, keep the bit out of the mouth until it is perfectly healed.

POLL-EVIL.

This is a disease the mule more than all other animals is subject to. This is more particularly so with those brought into the service of the Government unbroken.

It will be very easily seen that the necessary course of training, halter-breaking, &c., will expose them to many of the causes of this disease. Aside from this, the inhuman treatment of teamsters, and others who have charge of them, frequently produces it in its worst form. It begins with an ulcer or sore at the junction where the head and neck join; and from its position, more than any other cause, is very difficult to heal. The first thing to be done, when the swelling appears, is to use hot fomentations. If these are not at hand, use cold water frequently. Keep the bridle and halter from the parts. In case inflammation cannot be abated, and ulceration takes place, the only means to effect a cure, with safety and certainty, is by the use of the seton. This should be applied only by a hand well skilled in the use of it. The person should also well understand the anatomy of the parts, as injuries committed with the seton-needle, in those parts, are often more serious and more difficult of cure than the disease caused by the first injury.

FISTULA.

This is a disease the mule is more subject to than any other animal in Government use. And this, on account of his being used as a beast of burden by almost all nations and classes of people, and because he is the worst cared for. Fistula is the result of a bruise. Some animals have been known to produce it by rolling on stones and other hard substances. It generally makes its appearance first in the way of a rise or swelling where the saddle has been allowed to press too hard on the withers, and especially when the animal has high and lean ones. As the animal becomes reduced in flesh, the withers, as a matter of course, are more exposed and appear higher, on account of the muscle wasting from each side of the back-bone. This, under the saddle, can be remedied to a great extent, by adding an additional fold to the saddle blanket, or in making the pad of the saddle high enough to keep it from the withers. In packing with the pack-saddle this is more difficult, as the weight is generally a dead, heavy substance, and as the animal steps low or high, the pack does the same. Much, however, might be done by care in packing, to prevent injury to the withers and bruising of the back-bone. When the withers begin to swell and inflammation sets in, or a tumor begins to form, the whole may be driven away and the fistula scattered or avoided by frequent or almost constant applications of cold water--the same as is recommended in poll-evil. But if, in despite of this, the swelling should continue or become larger, warm fomentations, poultices, and stimulating embrocations should be applied, in order to bring the protuberance to its full formation as soon as possible. When full, a seton should be passed, by a skillful hand, from the top to the bottom of the tumor, so that all the pus may have free access of escape. The incision should be kept free until all the matter has escaped and the wound shows signs of healing. The after treatment must be similar to that recommended in the case of poll-evil. The above treatment, if properly administered, will in nearly all cases of *fistula* effect a cure.

COLLAR-GALLS.

Sore necks, saddle-galls, and stilfasts, are a species of injury and sore, which are in many cases very difficult of cure, especially saddle-galls on mules that have to be ridden every day. One of the best remedies for saddle gall is to heighten the saddle up as much as possible, and bathe the back with cold water as often as an opportunity affords. In many cases this will drive the fever away and scatter the trouble that is about to take place. This, however, does not always scatter, for the trouble will often continue, a root forming in the center of what we call the saddle-gall. The edges of this will be clear, and the stilfast hold only by the root. I have had many cases of this kind occur with the mule, both on his back and neck, mostly caused on the latter part by the collar being too loose. And I have found but one way to effectually cure them. Some persons advise cutting, which I think is too tedious and painful to the animal. My advice is to take a pair of pincers, or forceps of any kind, and pull it out. This done, bathe frequently with cold water, and keep the collar or saddle as much free of the sore as possible. This will do more towards relieving the animal and healing the injury than all the medicine you can give. A little soothing oil, or grease free from salt, may be rubbed lightly on the parts as they begin to heal. This is a very simple but effective remedy.

THRUSH.

This is another trouble with which the mule is afflicted. Cut away the parts of the frog that seem to be destroyed, clean the parts well with castile-soap, and apply muriatic acid. If you have not this at hand, a little tar mixed with salt, and placed on oakum or tow, and applied, will do nearly as well. Apply this every day, keeping the parts well dressed, and the feet according to directions in shoeing, and the trouble will soon disappear.

CHEST FOUNDERS.

Mules are not subject to this disease. Some persons assert that they are, but it is a mistake. These persons mistake for founder in the chest what is nothing more than a case of contraction of the feet. I have repeatedly seen veterinary surgeons connected with the army, on being asked what was the trouble with a mule, look wise, and declare the complaint chest founder, swelling of the shoulders, &c. I was inclined to put some faith in the wisdom of these gentlemen, until Doctor Braley, chief veterinary surgeon of the department of Washington, produced the most con-

vincing proofs that it was almost an impossibility for these animals to become injured in the shoulder. When mules become sore in front, look well to their feet, and
in nine cases out of ten, you will find the cause of the trouble there. In very many
cases a good practical shoer can remove the trouble by proper paring and shoeing.

BLEEDING.

It was always a subject of inquiry with me, who originated the system of bleeding; and why it was that all kinds of doctors and physicians persist in taking the
stream of life itself from the system in order to preserve life. In the case of General
Washington, which I copy from the ***Independent Chronicle*** of Boston, January 6,
1800, the editor, using "James Craik, physician, and Elisha C. Dick, physician," as
authority, states that a bleeder was procured in the neighborhood, who took from
the General's arm from twelve to fourteen ounces of blood, in the morning; and
in the afternoon of the same day was bled copiously twice. More than that, it was
agreed upon by these same enlightened doctors, to try the result of another blood-
letting, by which thirty two ounces more was drawn. And, wonderful as it may
seem to the intelligent mind at this day, they state that all this was done without
the slightest alleviation of the disease. The world has become more wise now, and
experience has shown how ridiculous this system of bleeding was. What is true in
regard to the human system is also true in regard to the animal. There are some
extreme cases in which I have no doubt moderate bleeding might render relief. But
these cases are so few that it should only be suffered to be done by an experienced,
careful, and skillful person. My advice is, avoid it in all cases where you can.

COLIC.

The mule is quite subject to this complaint. It is what is commonly known as
belly-ache. Over doses of cold water will produce it. There is nothing, however, so
likely to produce it in the mule as changes of grain.

Musty corn will also produce it, and should never be given to animals. I recollect, in 1856, when I was in New Mexico, at Fort Union, we had several mules die
from eating what is termed Spanish or Mexican corn, a small blue and purplish
grain. It was exceedingly hard and flinty, and, in fact, more like buckshot than
grain. We fed about four quarts of this to the mule, at the first feed. The result was,
they swelled up, began to pant, look round at their sides, sweat above the eyes and
at the flanks. Then they commenced to roll, spring up suddenly, lie down again, roll

and try to lie on their backs. Then they would spring up, and after standing a few seconds, fall down, and groan, and pant. At length they would resign themselves to what they apparently knew to be their fate, and die. And yet, singular as it may seem, the animal could be accustomed to this grain by judicious feeding at first.

We did not know at that time what to give the animal to relieve or cure him; and the Government lost hundreds of valuable animals through our want of knowledge. Whenever these violent cases appear, get some common soap, make a strong suds and drench the mule with it. I have found in every case where I used it that the mule got well. It is the alkali in the soap that neutralizes the gases. There is another good receipt, and it is generally to be found in camp. Take two ounces of saleratus, put it into a pint of water, shake well, and then drench with the same. Above all things, keep whisky and other stimulants away, as they only serve to aggravate the disease.

PHYSICKING.

This is another of those imaginary cures resorted to by persons having charge of mules. Very many of these persons honestly believe that it is necessary to clean the animal out every spring with large doses of poisonous and other truck. This, they say, ought to be given to loosen the hide, soften the hair, &c. In my opinion it does very little good. If his dung gets dry, and his hair hard and crispy, give him bran mashes mixed with his grain, and a teaspoonful of salt at each feed. If there is grass, let him graze a few hours every day. This will do more towards softening his coat and loosening his bowels than any thing else. When real disease makes its appearance, it is time to use medicines; but they should be applied by some one who thoroughly understands them.

STRINGHALT.

This sometimes occurs in the mule. It is a sudden, nervous, quick jerk of either or both of the hind legs. In the mule it frequently shows but little after being worked an hour or so. It is what I regard as unsoundness, and a mule badly affected with it is generally of but little use. It is often the result of strains, caused by backing, pulling and twisting, and heavy falls. You can detect it in its slightest form by turning the animal short around to the right or to the left. Turn him in the track he stands in, as near as possible, and then back him. If he has it, one of these three ways will develop its symptoms. There are a great many opinions as to the soundness or unsoundness of an animal afflicted with this complaint. If I had now a good animal afflicted with it, the pain caused to my feelings by looking at it would be a serious drawback.

CRAMP.

I have now under my charge several mules that are subject to this complaint. It does not really injure them for service, but it is very disagreeable to those having them in charge. It frequently requires from half an hour to two hours to get them rubbed so as the blood gets to its proper circulation, and to get them to walk without dragging their legs. In cases where they are attacked violently, they will appear to lose all use of their legs. I have known cases when a sudden stroke with a light piece of board, so as to cause a surprise, would drive it away. In other cases sudden application of the whip would have the same effect.

SPAVIN.

It is generally believed that the mule does not inherit this disease. But this is not altogether true. Small, compact mules, bred after the jack, are indeed not subject to it. On the contrary, large mules, bred from large, coarse mares, are very frequently afflicted with it. The author has under his charge at the present time quite a number of those kind of mules, in which this disease is visible. At times, when worked hard, they are sore and lame. The only thing to be recommended in this case is careful treatment, and as much rest at intervals as it is possible to give them. Hand rubbing and application of stimulant liniments, or tincture of arnica, is about all that can be done. The old method of firing and blistering only puts the animal to torture and the owner to expense. A cure can never be effected through it, and therefore should never be tried.

RINGBONE.

These appear on the same kind of large, bony mules as referred to in cases of spavin, and are incurable. They can, however, be relieved by the same process as recommended in spavin. Relief can also be afforded by letting the heels of the affected feet grow down to considerable length, or shoeing with a high-heeled shoe, and thus taking the weight or strain off the injured parts. The only way to make the best use or an animal afflicted with this disease, is to abandon experiments to effect a cure, as they will only be attended with expense and disappointment.

MANGE.

Mules are subject to this disease when kept in large numbers, as in the army. This is peculiarly a cuticle disease, like the itch in the human system, and yields to the same course of treatment. A mixture of sulphur and hog's lard, one pint of the latter to two of the former. Rub the animal all over, then cover with a blanket. After standing two days, wash him clean with soft-soap and water. After this process has been gone through, keep the animal blanketed for a few days, as he will be liable to take cold. Feed with bran mashes, plenty of common salt, and water. This will relieve the bowels all that is necessary, and can scarcely fail of effecting a cure. Another method, but not so certain in its effect, is to make a decoction of tobacco, say about one pound of the stems to two gallons of water, boiled until the strength is extracted from the weed, and when cool enough, bathe the mule well with it from head to foot, let him dry off, and do not curry him for a day or two. Then curry him well, and if the itching appear again, repeat the bathing two or three times, and it will produce a cure. The same treatment will apply in case of lice, which frequently occurs where mules are kept in large numbers. Mercury should never be used in any form, internally or externally, on an animal so much exposed as the mule.

GREASE-HEEL.

Clean the parts well with castile-soap and warm water. As soon as you have discovered the disease, stop wetting the legs, as that only aggravates it, and use ointment made from the following substances: Powdered charcoal, two ounces; lard or tallow, four ounces; sulphur, two ounces. Mix them well together, then rub the ointment in well with your hand on the affected parts. If the above is not at hand, get gunpowder, some lard or tallow, in equal parts, and apply in the same manner. If the animal be poor, and his system need toning up, give him plenty of nourishing food, with bran mash mixed plentifully with the grain. Add a teaspoonful of salt two or three times a day, as it will aid in keeping the bowels open. If the stable bottoms, or floors, or yards are filthy, see that they are properly cleaned, as filthiness is one of the causes of this disease. The same treatment will apply to scratches, as they are the same disease in a different form.

To avoid scratches and grease-heel during the winter, or indeed at any other season, the hair on the mule's heels should never be cut. Nor should the mud, in winter season, be washed off, but allowed to dry on the animal's legs, and then rubbed off with hay or straw. This washing, and cutting the hair off the legs, leave them without any protection, and is, in many cases, the cause of grease-heel and scratches.

SHOES, SHOEING, AND THE FOOT.

The foot, its diseases, and how to shoe it properly, is a subject much discussed among horsemen. Nearly every farrier and blacksmith has a way of his own for curing diseased feet, and shoeing. No matter how absurd it may be, he will insist that it has merits superior to all others, and it would be next to impossible to convince him of his error. Skillful veterinarians now understand perfectly all the diseases peculiar to the foot, and the means of effecting a cure. They understand, also, what sort of shoe is needed for the feet of different animals. Latterly number of shoes have been invented and patented, all professing to be exactly what is wanted to relieve and cure diseased feet of all kinds. One man has a shoe he calls "*concave*," and says it will cure contraction, corns, thrush, quarter-crack, toe-crack, &c., &c. But when you come to examine it closely, you will find it nothing more than a nicely dressed piece of iron, made almost in the shape of a half moon. After a fair trial, however, it will be found of no more virtue in curing diseases or relieving the animal than the ordinary shoe used by a country smithy. Another inventive genius springs up and asserts that he has discovered a shoe that will cure all sorts of diseased feet; and brings at least a bushel basket full of letters from persons he declares to be interested in the horse, confirming what he has said of the virtues of his shoe. But a short trial of this wonderful shoe only goes to show how little these persons understand the whole subject, and how easy a matter it is to procure letters recommending what they have invented.

Another has a "specific method" for shoeing, which is to cut away the toe right in the center of the foot, cut away the bars on the inside of the foot, cut and clean away all around on the inside of the hoof, then to let the animal stand on a board floor, so that his feet would be in the position a saucer would represent with one piece broken out at the front and two at the back. This I consider the most inhuman method in the art of shoeing. Turn this saucer upside down and see how little pressure it would bear, and you will have some idea of the cruelty of applying this "specific method." Sometimes bar-shoes and other contrivances are used, to keep the inside of the foot from coming down. But why do this? Why not get at once a shoe adapted to the spreading of the foot. Tyrell's shoe for this purpose is the best I have yet seen. We have used it in the Government service for two years, and experience has taught me that it has advantages that ought not to be overlooked. But even this shoe may be used to disadvantage by ignorant hands. Indeed, in the hands of a blacksmith who prefers "his own way," some kinds of feet may be just as badly injured by it as others are benefited. The United States Army affords the largest field for gaining practical knowledge concerning the diseases, especially of the feet, with which horses and mules are afflicted. During the late war, when so little care was given to animals in the field, when they were injured in every conceivable manner, and by all sorts of accidents, the veterinary found a field for study such as has never been opened before.

Experience has taught me, that common sense is one of the most essential things in the treatment of a horse's foot. You must remember that horses' feet differ as well as men's, and require different treatment, especially in shoeing. You must shoe the foot according to its peculiarity and demands, not according to any specific "system of shoe." Give the ground surface a level bearing, let the frog come to the ground, and the weight of the mule rest on the frog as much as any other part of the foot. If it project beyond the shoe, so much the better. That is what it was made for, and to catch the weight on an elastic principle. Never, under any circumstances, cut it away. Put two nails in the shoe on each side, and both forward of the quarters, and one in the toe, directly in front of the foot. Let those on the sides be an inch apart, then you will be sure not to cut and tear the foot. Let the nails and nail-holes be small, for they will then aid in saving the foot. It will still further aid in saving it by letting the nails run well up into the hoof, for that keeps the shoe steadier on the foot. The hoof is just as thick to within an inch of the top, and is generally sounder, and of a better substance, than it is at the bottom. Keep the first reason for shoeing apparent in your mind always--that you only shoe your mule because his feet will not stand the roads without it. And whenever you can, shoe him with a shoe exactly the shape of his foot. Some blacksmiths will insist on a shoe, and then cutting and shaping the foot to it. The first or central surface of the hoof, made hard by the animal's own peculiar way of traveling, indicates the manner in which he should be shod. All the art in the world cannot improve this, for it is the model prepared by nature. Let the shoes be as light as possible, and without calks if it can be afforded, as the mule always travels unsteady on them. The Goodenough shoe is far superior to the old calked shoe, and will answer every purpose where holding is necessary. It is also good in mountainous countries, and there is no danger of the animal calking himself with it. I have carefully observed the different effect of shoes, while with troops on the march. I accompanied the Seventh Infantry, in 1858, in its march to Cedar Valley, in Utah, a distance of fourteen hundred miles, and noticed that scarcely a man who wore regulation shoes had a blister on his feet, while the civilians, who did not, were continually falling out, and dropping to the rear, from the effects of narrow and improper shoes and boots. The same is the case with the animal. The foot must have something flat and broad to bear on. The first care of those having charge of mules, should be to see that their feet are kept in as near a natural state as possible. Then, if all the laws of nature be observed, and strictly obeyed, the animal's feet will last as long, and be as sound in his domestic state as he would be in a state of nature.

The most ordinary observer will soon find that the outer portion or covering of the mule's foot possesses very little animal life, and has no sensibility, like the hair or covering of the body. Indeed, the foot of the horse and mule is a dense block of horn, and must therefore be influenced and governed by certain chemical laws, which control the elements that come in contact with it. Hence it was that the feet of these animals was made to bear on the hard ground, and to be wet naturally every time the horse drank. Drought and heat will contract and make hard and brittle the substance of which the feet is composed; while on the other hand cooling and moisture will expand it, and render it pliable and soft. Nature has provided everything necessary to preserve and protect this foot, while the animal is in a natural state; but when brought into domestic use, it requires the good sense of man, whose servant he is, to artificially employ those means which nature has provided, to keep it perfectly healthy.

When, then, the foot is in a healthy state, wet it at least twice a day; and do not be content with merely throwing cold water on the outside, for the foot takes in very little if any moisture through the wall. In short, it absorbs moisture most through the frog and sole, particularly in the region where the sole joins the wall. This, if covered by a tight shoe, closes the medium, and prevents the proper supply. Horses that are shod should be allowed to stand in moist places as much as possible.

Use clay or loam floors, especially if the horse has to stand much of his time. Stone or brick is the next best, as the foot of the animal will absorb moisture from either of these. Dry pine planks are the very worst, because they attract moisture from the horse's foot. Where animals have to stand idle much of the time, keep their feet well stuffed with cow manure at night. That is the best and cheapest preservative of the feet that you can use.

ADVICE TO BLACKSMITHS.

Let me enjoin you, for humanity's sake, that when you first undertake to shoe a young animal, you will not forget the value of kind treatment. Keep its head turned away from the glaring fire, the clinking anvil, &c., &c. Let the man whom he has been accustomed to, the groom or owner, stand at his head, and talk to him kindly. When you approach him for the first time, let it be without those implements you are to use in his shoeing. Speak to him gently, then take up his foot. If he refuse to let you do this, let the person having him in charge do it. A young animal will allow this with a person he is accustomed to, when he will repel a stranger. By treating him kindly you can make him understand what is wanted; by abusing him you will only frighten him into obstinacy. When you have got the animal under perfect subjection, examine the foot carefully, and you will find the heels, at the back part of the frog, entirely free from that member, which is soft and spongy. When the foot is down, resting on the ground, grasp the heels in your strong hand, press them inwards towards the frog, and you will immediately find that they will yield. You will then see that what yields so easily to the mere pressure of the hand will expand and spread out when the weight of the body is thrown on it. This should give you an idea of what you have to do in shoeing that foot, and your practical knowledge should stand you well in an argument with any of those "learned professors," who declare the foot of the mule does not expand or contract. In truth it is one of its necessary conditions. After being a long time badly shod, nearly or all of this necessary principle of the foot will be lost. You should therefore study to preserve it. And here let me give you what little aid experience has enabled me to do. You will observe the ground surface of the foot, no matter how high the arch may be, to be at least half an inch wide, and sometimes more than an inch, with the heels spread out at the outside quarter. Do not cut away this important brace. It is as necessary to the heel of the animal, to guard him against lateral motion, on which the whole of the above structure depends, as the toes are to the human being. Curve the outside of the shoe nearly to fit the foot, and you will find the inside heel a little straighter, especially if the animal be narrow-breasted, and the feet stand close together. Nature has provided this safeguard to prevent its striking the opposite leg. After the shoe is prepared to fit the foot, as I have before described, rasp the bottom level--it will be found nearly so. Do not put a knife to the sole or the frog. The sole of the foot, remember, is its life, and the frog its defender. In punching the shoe, two nail-holes on a side, on a foot like this, are sufficient to hold on a shoe. Three may be used, if set in their proper places, without injury to the foot. Practice will teach you that any more nailing than this is unnecessary. I have used two nails on a side on an animal with not the best of a foot, and very high action, and he has worn them entirely out without throwing either of them off. Previous to punching the shoe, observe the grain of the foot. It will be seen that the fibres of the hoof run from the top of the foot, or coronary border, towards the toe, in most feet, at an angle of about forty-five degrees. It will be plain, then, that if the nails are driven with the grain of the horn, they will drive much easier, and hold better, and be less liable to cut and crack the fibers.

Another benefit can be derived from this process of nailing. When the foot comes to the ground, the nails act as a brace to keep the foot from slipping forward off the shoe. This renders that very ingenious foot destroyer, the toe-clip, unnecessary. Then, in punching the shoe, hold the top of the pritchell toward the heel of the shoe, so as to get the hole in the shoe on an angle with the grain of the hoof. Punch the holes large enough, so that the nails will not bind in the shoe, nor require unnecessary hammering or bruising of the foot to get them up to their proper place. Prepare the nails well, point them thin and narrow; and, as I have said before, use as small a nail as possible.

When you proceed to nail on the shoe, take a slight hold at the bottom, so as to be sure that the nail starts in the wall of the foot instead of the sole. Let it come out as high up as possible. You need not be afraid of pricking with nails set in this way, as the wall of the foot is as thick, until you get within half an inch of the top, as it is where you set the nail. Nails driven in this way injure the feet less, hold on longer, and are stronger than when driven in any other way. If you have any doubt of this, test it in this manner: when you take off an old shoe to set a new one, and cut the clinches (which should be done in all cases), you will find the old nail and the clinches not started up; and in drawing the nail out you will also find the foot not slipped or cracked; and that the horn binds the nail until it is entirely drawn out. Indeed, I have known the hole to almost close as the nail left it.

Set the two front nails well towards the toe, so as not to be more than two inches apart when measured across the bottom of the foot. Let the next two divide the distance from that to the heel, so as to leave from two to two and a half inches free of nails, as the form of the foot may allow. Lastly, before nailing on the shoe, and while it is cold on the anvil, strike the surface that comes next to the foot on the outside, a few blows with the hammer, right across the heels, and see also that the outside of the heels is a shade lowest, so that the animal in throwing his weight upon them will spread out, and not pinch in his feet.

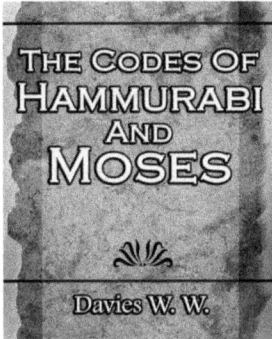

The Codes Of Hammurabi And Moses
W. W. Davies

QTY

The discovery of the Hammurabi Code is one of the greatest achievements of archaeology, and is of paramount interest, not only to the student of the Bible, but also to all those interested in ancient history...

Religion   **ISBN:** *1-59462-338-4*   **Pages:132**
MSRP $12.95

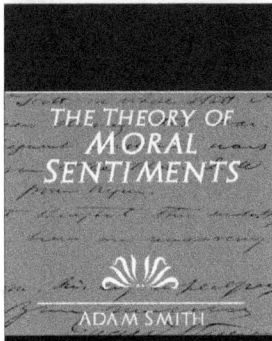

The Theory of Moral Sentiments
Adam Smith

QTY

This work from 1749. contains original theories of conscience amd moral judgment and it is the foundation for systemof morals.

Philosophy **ISBN:** *1-59462-777-0*   **Pages:536**
MSRP $19.95

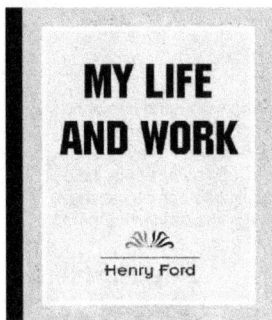

Jessica's First Prayer
Hesba Stretton

QTY

In a screened and secluded corner of one of the many railway-bridges which span the streets of London there could be seen a few years ago, from five o'clock every morning until half past eight, a tidily set-out coffee-stall, consisting of a trestle and board, upon which stood two large tin cans, with a small fire of charcoal burning under each so as to keep the coffee boiling during the early hours of the morning when the work-people were thronging into the city on their way to their daily toil...

Pages:84

Childrens   **ISBN:** *1-59462-373-2*   *MSRP $9.95*

My Life and Work
Henry Ford

QTY

Henry Ford revolutionized the world with his implementation of mass production for the Model T automobile. Gain valuable business insight into his life and work with his own auto-biography... "We have only started on our development of our country we have not as yet, with all our talk of wonderful progress, done more than scratch the surface. The progress has been wonderful enough but..."

Pages:300

Biographies/   **ISBN:** *1-59462-198-5*   *MSRP $21.95*

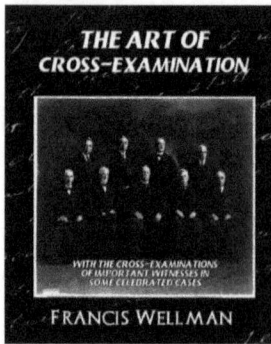

The Art of Cross-Examination
Francis Wellman

QTY

I presume it is the experience of every author, after his first book is published upon an important subject, to be almost overwhelmed with a wealth of ideas and illustrations which could readily have been included in his book, and which to his own mind, at least, seem to make a second edition inevitable. Such certainly was the case with me; and when the first edition had reached its sixth impression in five months, I rejoiced to learn that it seemed to my publishers that the book had met with a sufficiently favorable reception to justify a second and considerably enlarged edition. ..

Pages:412

Reference   ISBN: *1-59462-647-2*   *MSRP $19.95*

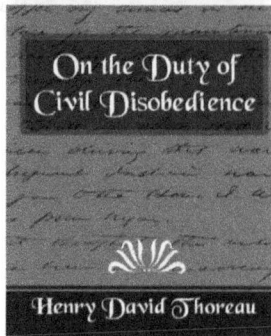

On the Duty of Civil Disobedience
Henry David Thoreau

QTY

Thoreau wrote his famous essay, On the Duty of Civil Disobedience, as a protest against an unjust but popular war and the immoral but popular institution of slave-owning. He did more than write—he declined to pay his taxes, and was hauled off to gaol in consequence. Who can say how much this refusal of his hastened the end of the war and of slavery ?

Law   ISBN: *1-59462-747-9*   **Pages:48**

MSRP $7.45

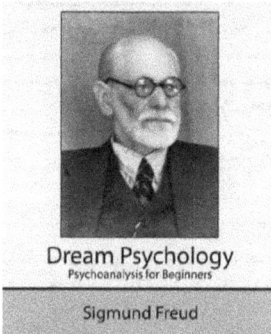

Dream Psychology Psychoanalysis for Beginners
Sigmund Freud

QTY

Sigmund Freud, born Sigismund Schlomo Freud (May 6, 1856 - September 23, 1939), was a Jewish-Austrian neurologist and psychiatrist who co-founded the psychoanalytic school of psychology. Freud is best known for his theories of the unconscious mind, especially involving the mechanism of repression; his redefinition of sexual desire as mobile and directed towards a wide variety of objects; and his therapeutic techniques, especially his understanding of transference in the therapeutic relationship and the presumed value of dreams as sources of insight into unconscious desires.

Pages:196

Psychology   ISBN: *1-59462-905-6*   *MSRP $15.45*

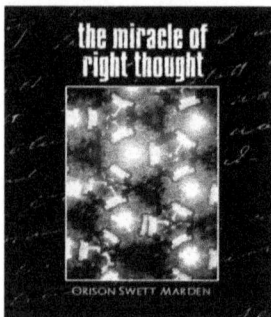

The Miracle of Right Thought
Orison Swett Marden

QTY

Believe with all of your heart that you will do what you were made to do. When the mind has once formed the habit of holding cheerful, happy, prosperous pictures, it will not be easy to form the opposite habit. It does not matter how improbable or how far away this realization may see, or how dark the prospects may be, if we visualize them as best we can, as vividly as possible, hold tenaciously to them and vigorously struggle to attain them, they will gradually become actualized, realized in the life. But a desire, a longing without endeavor, a yearning abandoned or held indifferently will vanish without realization.

Pages:360

Self Help   ISBN: *1-59462-644-8*   *MSRP $25.45*

QTY

The Rosicrucian Cosmo-Conception Mystic Christianity by *Max Heindel*　ISBN: *1-59462-188-8*　**$38.95**
The Rosicrucian Cosmo-conception is not dogmatic, neither does it appeal to any other authority than the reason of the student. It is: not controversial, but is: sent forth in the, hope that it may help to clear..　New Age/Religion Pages 646

Abandonment To Divine Providence by *Jean-Pierre de Caussade*　ISBN: *1-59462-228-0*　**$25.95**
"The Rev. Jean Pierre de Caussade was one of the most remarkable spiritual writers of the Society of Jesus in France in the 18th Century. His death took place at Toulouse in 1751. His works have gone through many editions and have been republished...　Inspirational/Religion Pages 400

Mental Chemistry by *Charles Haanel*　ISBN: *1-59462-192-6*　**$23.95**
Mental Chemistry allows the change of material conditions by combining and appropriately utilizing the power of the mind. Much like applied chemistry creates something new and unique out of careful combinations of chemicals the mastery of mental chemistry...　New Age/Business Pages 354

The Letters of Robert Browning and Elizabeth Barret Barrett 1845-1846 vol II　ISBN: *1-59462-193-4*　**$35.95**
by *Robert Browning* and *Elizabeth Barrett*　Biographies Pages 596

Gleanings In Genesis (volume I) by *Arthur W. Pink*　ISBN: *1-59462-130-6*　**$27.45**
Appropriately has Genesis been termed "the seed plot of the Bible" for in it we have, in germ form, almost all of the great doctrines which are afterwards fully developed in the books of Scripture which follow...　Religion/Inspirational Pages 420

The Master Key by *L. W. de Laurence*　ISBN: *1-59462-001-6*　**$30.95**
In no branch of human knowledge has there been a more lively increase of the spirit of research during the past few years than in the study of Psychology, Concentration and Mental Discipline. The requests for authentic lessons in Thought Control, Mental Discipline and...　New Age/Business Pages 422

The Lesser Key Of Solomon Goetia by *L. W. de Laurence*　ISBN: *1-59462-092-X*　**$9.95**
This translation of the first book of the "Lernegton" which is now for the first time made accessible to students of Talismanic Magic was done, after careful collation and edition, from numerous Ancient Manuscripts in Hebrew, Latin, and French...　New Age/Occult Pages 92

Rubaiyat Of Omar Khayyam by *Edward Fitzgerald*　ISBN:*1-59462-332-5*　**$13.95**
Edward Fitzgerald, whom the world has already learned, in spite of his own efforts to remain within the shadow of anonymity, to look upon as one of the rarest poets of the century, was born at Bredfield, in Suffolk, on the 31st of March, 1809. He was the third son of John Purcell...　Music Pages 172

Ancient Law by *Henry Maine*　ISBN: *1-59462-128-4*　**$29.95**
The chief object of the following pages is to indicate some of the earliest ideas of mankind, as they are reflected in Ancient Law, and to point out the relation of those ideas to modern thought.　Religiom/History Pages 452

Far-Away Stories by *William J. Locke*　ISBN: *1-59462-129-2*　**$19.45**
"Good wine needs no bush, but a collection of mixed vintages does. And this book is just such a collection. Some of the stories I do not want to remain buried for ever in the museum files of dead magazine-numbers an author's not unpardonable vanity..."　Fiction Pages 272

Life of David Crockett by *David Crockett*　ISBN: *1-59462-250-7*　**$27.45**
"Colonel David Crockett was one of the most remarkable men of the times in which he lived. Born in humble life, but gifted with a strong will, an indomitable courage, and unremitting perseverance...　Biographies/New Age Pages 424

Lip-Reading by *Edward Nitchie*　ISBN: *1-59462-206-X*　**$25.95**
Edward B. Nitchie, founder of the New York School for the Hard of Hearing, now the Nitchie School of Lip-Reading, Inc, wrote "LIP-READING Principles and Practice". The development and perfecting of this meritorious work on lip-reading was an undertaking...　How-to Pages 400

A Handbook of Suggestive Therapeutics, Applied Hypnotism, Psychic Science　ISBN: *1-59462-214-0*　**$24.95**
by *Henry Munro*　Health/New Age/Health/Self-help Pages 376

A Doll's House: and Two Other Plays by *Henrik Ibsen*　ISBN: *1-59462-112-8*　**$19.95**
Henrik Ibsen created this classic when in revolutionary 1848 Rome. Introducing some striking concepts in playwriting for the realist genre, this play has been studied the world over.　Fiction/Classics/Plays 308

The Light of Asia by *sir Edwin Arnold*　ISBN: *1-59462-204-3*　**$13.95**
In this poetic masterpiece, Edwin Arnold describes the life and teachings of Buddha. The man who was to become known as Buddha to the world was born as Prince Gautama of India but he rejected the worldly riches and abandoned the reigns of power when...　Religion/History/Biographies Pages 170

The Complete Works of Guy de Maupassant by *Guy de Maupassant*　ISBN: *1-59462-157-8*　**$16.95**
"For days and days, nights and nights, I had dreamed of that first kiss which was to consecrate our engagement, and I knew not on what spot I should put my lips..."　Fiction/Classics Pages 240

The Art of Cross-Examination by *Francis L. Wellman*　ISBN: *1-59462-309-0*　**$26.95**
Written by a renowned trial lawyer, Wellman imparts his experience and uses case studies to explain how to use psychology to extract desired information through questioning.　How-to/Science/Reference Pages 408

Answered or Unanswered? by *Louisa Vaughan*　ISBN: *1-59462-248-5*　**$10.95**
Miracles of Faith in China　Religion Pages 112

The Edinburgh Lectures on Mental Science (1909) by *Thomas*　ISBN: *1-59462-008-3*　**$11.95**
This book contains the substance of a course of lectures recently given by the writer in the Queen Street Hall, Edinburgh. Its purpose is to indicate the Natural Principles governing the relation between Mental Action and Material Conditions...　New Age/Psychology Pages 148

Ayesha by *H. Rider Haggard*　ISBN: *1-59462-301-5*　**$24.95**
Verily and indeed it is the unexpected that happens! Probably if there was one person upon the earth from whom the Editor of this, and of a certain previous history, did not expect to hear again...　Classics Pages 380

Ayala's Angel by *Anthony Trollope*　ISBN: *1-59462-352-X*　**$29.95**
The two girls were both pretty, but Lucy who was twenty-one who supposed to be simple and comparatively unattractive, whereas Ayala was credited, as her Bombwhat romantic name might show, with poetic charm and a taste for romance. Ayala when her father died was nineteen...　Fiction Pages 484

The American Commonwealth by *James Bryce*　ISBN: *1-59462-286-8*　**$34.45**
An interpretation of American democratic political theory. It examines political mechanics and society from the perspective of Scotsman James Bryce　Politics Pages 572

Stories of the Pilgrims by *Margaret P. Pumphrey*　ISBN: *1-59462-116-0*　**$17.95**
This book explores pilgrims religious oppression in England as well as their escape to Holland and eventual crossing to America on the Mayflower, and their early days in New England...　History Pages 268

QTY

The Fasting Cure *by Sinclair Upton*   ISBN: *1-59462-222-1*   **$13.95**
In the Cosmopolitan Magazine for May, 1910, and in the Contemporary Review (London) for April, 1910, I published an article dealing with my experiences in fasting. I have written a great many magazine articles, but never one which attracted so much attention... New Age/Self Help/Health Pages 164

Hebrew Astrology *by Sepharial*   ISBN: *1-59462-308-2*   **$13.45**
In these days of advanced thinking it is a matter of common observation that we have left many of the old landmarks behind and that we are now pressing forward to greater heights and to a wider horizon than that which represented the mind-content of our progenitors... Astrology Pages 144

Thought Vibration or The Law of Attraction in the Thought World   ISBN: *1-59462-127-6*   **$12.95**
by William Walker Atkinson   *Psychology/Religion Pages 144*

Optimism *by Helen Keller*   ISBN: *1-59462-108-X*   **$15.95**
Helen Keller was blind, deaf, and mute since 19 months old, yet famously learned how to overcome these handicaps, communicate with the world, and spread her lectures promoting optimism. An inspiring read for everyone... Biographies/Inspirational Pages 84

Sara Crewe *by Frances Burnett*   ISBN: *1-59462-360-0*   **$9.45**
In the first place, Miss Minchin lived in London. Her home was a large, dull, tall one, in a large, dull square, where all the houses were alike, and all the sparrows were alike, and where all the door-knockers made the same heavy sound... Childrens/Classic Pages 88

The Autobiography of Benjamin Franklin *by Benjamin Franklin*   ISBN: *1-59462-135-7*   **$24.95**
The Autobiography of Benjamin Franklin has probably been more extensively read than any other American historical work, and no other book of its kind has had such ups and downs of fortune. Franklin lived for many years in England, where he was agent... Biographies/History Pages 332

Name	
Email	
Telephone	
Address	
City, State ZIP	

☐ **Credit Card**   ☐ **Check / Money Order**

Credit Card Number	
Expiration Date	
Signature	

Please Mail to: Book Jungle
PO Box 2226
Champaign, IL 61825
or Fax to: 630-214-0564

ORDERING INFORMATION

web: *www.bookjungle.com*
email: *sales@bookjungle.com*
fax: *630-214-0564*
mail: *Book Jungle PO Box 2226 Champaign, IL 61825*
or PayPal *to sales@bookjungle.com*

Please contact us for bulk discounts

DIRECT-ORDER TERMS

**20% Discount if You Order
Two or More Books**
Free Domestic Shipping!
Accepted: Master Card, Visa,
Discover, American Express